A Travel Guide
to
Scientific Sites
of
The British Isles

A Travel Guide
to
Scientific Sites
of
The British Isles

A Guide to the People, Places
and Landmarks of Science

Charles Tanford
Jacqueline Reynolds

John Wiley & Sons

Chichester • New York • Brisbane • Toronto • Singapore

Other Wiley Editorial Offices

John Wiley & Sons, Inc., 605 Third Avenue,
New York, NY 10158-0012, USA

Jacaranda Wiley Ltd, 33 Park Road, Milton,
Queensland 4064, Australia

John Wiley & Sons (Canada) Ltd, 22 Worcester Road,
Rexdale, Ontario M9W 1L1, Canada

John Wiley & Sons (SEA) Pte Ltd, 37 Jalan Pemimpin #05-04,
Block B, Union Industrial Building, Singapore 2057

Library of Congress Cataloging-in-Publication Data

Tanford, Charles, 1921–
A travel guide to scientific sites of the British Isles:
a guide to the people, places and landmarks of
science/Charles Tanford, Jacqueline Reynolds.
p.   cm.
Includes indexes.
ISBN 0-471-95270-2
1. Science—Great Britain—History—Guidebooks.
2. Research institutes—Great Britain—History—
Guidebooks. 3. Scientists—Great Britain—Handbooks,
manuals, etc.
I. Reynolds, Jacqueline A. (Jacqueline Ann), 1930–
II. Title.
Q127.G436   1995
509.41—dc20   94–42176
                                                      CIP

British Library Cataloging in Publication Data

A catalogue record for this book is available from the British
Library

ISBN 0-471-95270-2

Typeset in 9/11pt Times from authors' disks by
Central Southern Typesetters, Eastbourne
Printed and bound in Great Britain by Bourne Press, Bournemouth

CONTENTS

PREFACE

The population of the British Isles is less than 0.2% that of the entire earth; yet this tiny fraction of human society is responsible for an enormous number of cultural advances in both the arts and the sciences. Public appreciation for the men and women of Britain and Ireland who wrote, painted, composed music, etc. is evident wherever one looks, but the recognition of explorers of nature is harder to find. We began this book with the idea that this situation could be rectified if only a little more publicity were given to the home-grown scientists who have been among the leaders in nearly every new advance, often laying the foundations of what would prove to be (and still are) principal fields of knowledge in the overall domain of natural science.

Since we like to travel and have noted the penchant of many others to do so (evidenced by the motoring mania on our roads), a travel guide to places associated with science seemed an obvious way to highlight our scientific heritage. We have accordingly divided the book in two parts. Part I describes some selected people and events that, in our opinion, "turned the world upside down" — landmarks in the march to understand the world around us. Part II directs the reader to places of various kinds — museums, homes, churches, memorials, laboratories — where the individuals from Part I together with many others (perhaps less well known, but not necessarily deservedly so) are remembered in some way. The people and places we have chosen are by no means all-inclusive, but we have attempted to cover a broad range of both basic and applied science, within the framework of our own particular preferences. Other authors may well have made different choices.

PRACTICAL NOTES

We assume the reader will have appropriate maps but have given specific directions in any case where a place may be a little hard to find. We have found tourist offices extremely helpful when we lost our way — sometimes they may also know about historic places that we may have missed. We have provided telephone numbers for public institutions and recommend checking on opening times since these sometimes change. Sadly, many of our smaller churches are kept locked except during normal services, so we have again provided information when available as to how to obtain access.

ACKNOWLEDGEMENTS

We are grateful to our editor, Geoffrey Farrell, for his encouragement and advice. Some parts of the text of this book have been adapted from an earlier one by us, *The Scientific Traveler*, also published by John Wiley & Sons. We particularly wish to thank all those custodians of archives and museums who have been most helpful to us not only in providing information, but in their enthusiasm for our project.

Most of the illustrations are the authors' own photographs. Sources for others are as follows: **5** All Saints' Church, Ockham; **24** Department of Physics, Cambridge University; **68** The Director of the Royal Institution; **80** Science Museum, London; **112** Bowood Estate, Calne; **140** English Heritage; **146** Jenner Museum, Berkeley; **216** Grantham Museum, Lincolnshire County Council; **234** Manchester City Council; **323** from *The Cambridge Encyclopedia of Earth Sciences*.

Permission is gratefully acknowledged for the use of the following photographs on the front cover: Architecture Association (St. Bartholomew's Hospital); Science Museum (Herschel's telescope); David Parker/Science Photo Library (Stonehenge); Sinclair Stammers/Science Photo Library (Ammonites).

Part I
Turning The World Upside Down

On 17 April 1993 a small public meeting was held in Edinburgh, entitled "Scotland's Uncelebrated Genius: James Clerk Maxwell 1831–1879". The venue was the house where Maxwell was born. What is astonishing about this event is its title: how can Maxwell be called uncelebrated? He is universally regarded by physicists as one of the greatest of their clan. Albert Einstein called his work "the most fruitful that physics has experienced since the time of Newton." He discovered that light is a form of electromagnetic radiation. His work constitutes the foundation for a huge part of what we prize as the essence of our advanced technology: telephone, television, etc. How is it possible that he can be uncelebrated in his native city?

It is, of course, true — not only in Edinburgh, but throughout the British Isles — that admirals and generals, novelists, politicians, and even musical composers of quite modest stature are honoured, while scientists tend to be pushed into the background. The fact of the matter is that Britain has been in the forefront of scientific discovery since the earliest days of western civilisation. Scientific research and speculation were practically a cottage industry for several centuries, extending to every county, often to the smallest hamlet. Clergymen preached science from the pulpit and often did their own bit of research on the side.

We have much to be proud of, locally as well as nationally, and a few selected episodes — some highlights of major scientific advances — serve to illustrate the point. Here, indeed, are people and events that turned the scientific world upside down.

THE BEGINNING

When we think of primitive peoples, we usually place them in a hostile world struggling to survive the elements and expending their energies on food gathering and protection from predators. But other, somewhat surprising activities are evident ranging from prehistoric cave paintings on the European continent to construction projects throughout the British Isles that required a mechanical skill we tend to associate with modern societies.

3

Along the west coast of Britain, from Orkney down to Cornwall, there are hundreds of stone circles or archaeological remains of stone circles, claimed by some to represent stone age calendar clocks, deliberately built, with stones intentionally placed in specified positions for unique recognition of midsummer day, for example. But this claim is controversial and there are others who argue that the circles had a more general ceremonial function, even if they might secondarily have been used as the community calendars. Stonehenge, 90 miles west of London, is the grandest and most famous of these circles, a massive monument standing alone on Salisbury Plain. This is the site to which virtually all research, speculation and public attention has been directed. But in Scotland, too, we find evidence of these early builders such as the Ring of Brogar and the nearby Stenness Stones on the island of Orkney. These are among the oldest such relics, built between 2500 and 3000 B.C., as compared to about 1600 B.C. for Stonehenge, and they are set in an area rich in other stone age sites.

Is this visual evidence of an ability to design and carry out construction projects of such magnitude in the absence of modern mechanical devices related to science as we know it? If one accepts the premise of G.E.R. Lloyd, the noted historian, that technological development (while perhaps involving no conscious theorising) demonstrates a highly developed ability to observe and to learn from experience, then, indeed, these early builders displayed two of the basic qualities necessary for any field of science.

PHILOSOPHERS

Early Western intellectual activity focused on the study of ancient Greek and Arabic manuscripts introduced into Europe by the Moors during their long occupation of the Iberian peninsula (711 until the fall of the last Moorish stronghold in Granada in 1491). But by the 13th century, in Britain as well as the rest of Europe, philosophers had begun to expound their own original views about the natural world and the means that should be used to study it. Their

William of Occam. Window in the church in Ockham, Surrey.

pronouncements may seem to us commonplace but in those times they were a radical departure from accepted ideas of scholarship. **Robert Grosseteste** (1175–1253) and **Roger Bacon** (1214–1294) were early leaders in advocating what might be called experimental science — the making of accurate observations of phenomena, the design of controlled experiments. And somewhat later **William of Ockham** (d. 1349) left us his philosophical "razor", "What can be done with fewer assumptions, is done in vain with more" — a scientific philosophy still taught to students today. A more widely known philosopher is **Francis Bacon** (1561–1626) who espoused the idea of inductive reasoning — namely, inferring a general law or principle from observation of particular instances. Despite his lasting fame, most modern scientists would find his namesake, Roger Bacon, with his emphasis on careful and multitudinous observations designed to eliminate untrue hypotheses more akin to their own practice in the laboratory.

These early natural philosophers set the stage for a shift in scientific emphasis from the premise that reasoning alone could answer our questions about ourselves and our world to a new enthusiasm for observation and experimentation. They considered all science to be within their bailiwick and investigated the earth and heavens with equal enthusiasm. Over the ensuing centuries, however, different branches of science began to emerge, culminating today in a fragmentation into multitudinous disciplines and sub-disciplines. The age of the polymath is gone.

HUMAN BIOLOGY AND MEDICINE

The study of human anatomy and physiology and concern with human diseases can be traced back to earliest recorded history, and by the time of the Middle Ages many great institutions had emerged on the Continent where these sciences were in full flower. Britons went there to study and assimilate the exciting new scholarship, among them **Thomas Linacre** (1460–1524) who was to found the Royal College of Physicians after his return from the University of

Padua. But information did not flow in just one direction. In Britain discoveries were made equal to and sometimes more important than those of our Continental neighbours.

Circulation of the Blood

William Harvey (1578–1657) said it all in a slim volume of 72 pages, _De motu cordis et sanguinis_, dedicated to King Charles I. The blood circulates, he reported — the same blood is used over and over again. It goes from the left side of the heart to all parts of the body, then returns to the right side of the heart. It goes from there to the lungs for contact with freshly breathed air and to the left side of the heart, where the cycle begins anew. These simple words were enough for a revolution, for they destroyed a tangled web of unbelievable misinformation — "incongruous, obscure, impossible" in Harvey's own words — and they exerted an influence far beyond Harvey's cardiovascular system. Almost all of animal physiology converges on the circulation of the blood and wild confusion about the latter inevitably engenders distorted ideas about everything else.

Most of the doctrine about blood circulation taught before Harvey was based on the writings attributed to Galen. Venous blood was thought to be _nourishment_ intended for one time use, made in the liver from digested food and distributed by the veins to other parts of the body — all the major organs were thought to be made and kept in repair by use of the actual substance of the blood. Only a part of the venous blood was thought to go to the heart and lungs, to be infused with life's _vital spirit_, somehow produced from the air we inhale. (Some authorities believed that the arteries actually contain nothing but air.) Harvey's Padua teacher, Fabrici, had observed the tiny valves in the veins of human limbs, which in fact make a shambles of the Galenic doctrine because they face the wrong way, but Fabrici was blinded by doctrinal prejudice and saw the valves as instruments for _fair distribution_ of nutritive elements.

Harvey had no special skills or special tools to demolish the false doctrine and to chart the true course of the flow of blood, simply accurate observations that anyone else could have made, coupled to a refreshingly open mind. Harvey supported his observational conclusions with a devastating numerical calculation. The capacity of the left ventricle is about two ounces, and the heart beats about 72 times a minute. That means as much as 144 ounces (9 pounds) of blood might be ejected into the arteries per minute. Where can all this blood come from? Where is it going? There is only one possible answer — it must be the same blood, used again and again.

Harvey was born in Kent to a relatively prosperous family and was educated at Cambridge and in Padua, where he studied under the famous anatomist, Fabrici. When he returned to England he went to London to set up a medical practice. He became a fellow of the College of Physicians and took an active part in its affairs; he became physician of St. Bartholomew's Hospital in 1609 and continued in that position for 35 years; and he served as personal physician to James I and Charles I. He and Charles developed friendship and genuine affection for one another, and, when civil war came, Harvey left London as part of the king's retinue. The king's execution in Whitehall in January 1649 must have been shocking and sickening for his medical friend — Harvey was seventy years old at the time.

Immunity Against Disease

Another revolution, a complete rethink about how animals and people survive in an environment of hostile organisms, arose in the small town of Berkeley in Gloucestershire. **Edward Jenner** (1749–1823), the son of the local vicar, was orphaned at the age of five, but his elder brother Stephen saw to it that he had an excellent education. Edward was apprenticed to a local surgeon at the age of fourteen and then spent several years in London with the prominent surgeon and anatomist John Hunter. While in London, Jenner also worked with Sir Joseph Banks, Captain Cook's naturalist. Jenner's job was to preserve and arrange zoological specimens brought back from Cook's first voyage.

Jenner started medical practice in his native Berkeley in 1773, and we must suppose that he approached it with the same intelligent curiosity that he applied to his observations of cuckoos and hedgehogs, the topics of his first two published papers. His famous work was on the more serious subject of smallpox, one of the great scourges of the time.

Jenner was intrigued by the immunity of milkmen and milkmaids to smallpox, a resistance seemingly related to previous contraction of a "pox" disease of the teats of milk cows which was communicable to human beings. He reasoned that it might be possible to protect anyone from smallpox by deliberately infecting them with the less virulent cowpox. Twenty years of meticulous study lay ahead before he summoned the courage to perform the crucial experiment, but then it proved a huge success. A young boy inoculated with extract from a cowpox pustule proved immune to subsequent inoculation with a similar extract from a smallpox victim.

This small-town practitioner published his results in a slim volume at his own expense, and worldwide reaction was instantaneous. The "vaccine" (so named after the Latin word _vacca_ for "cow") could be preserved as a dry powder, and soon Jenner was busy sending samples to all corners of the globe. He was voted large grants by the British parliament, and a national programme of vaccination was begun. Though Britain and France were at war, Napoleon had a medal struck in honour of Jenner and made vaccination compulsory in the French army. Other national programmes followed and smallpox has recently been declared eradicated, even from the remotest corner of the earth.

Antiseptic Surgery

Joseph Lister (1827–1912), son of a London wine dealer, is another of the great figures in the history of medicine. He pioneered the introduction of antiseptic practices into surgery, doing so in the face of strong opposition and ridicule. This may at first seem remote from basic science, but in fact the contrary is true because the germ theory of disease was at the heart of the matter. It was Louis Pasteur who convinced the world that microorganisms exist and cause fermentation and putrefaction, but Lister was one of the first to recognise microbes as a _cause of disease_, an idea which his professional colleagues were initially unable to accept. The

_9_

underlying cause for dispute was the great despot of German chemistry, Justus Liebig, who died in 1873, but whose rejection of any hint of "vitalism" in what he thought to be purely chemical processes (putrefaction among them) retained an influence for several years after his death.

Lister received his medical degree in London in 1852 and then went to Edinburgh to spend a month with James Syme, considered the most original surgeon of his day. As it happened, Syme and Lister took an instant liking to each other and Lister married Syme's oldest daughter Agnes. Lister remained in Scotland for over twenty years, first in Glasgow and then in Edinburgh as Syme's successor. Mortality rates of surgical patients were appallingly high at this time, even for simple amputations. Patients died like flies from blood poisoning and the especially dreaded hospital gangrene. The standard explanation was (*à la* Liebig) that chemical oxidative degradations were responsible and scrupulous avoidance of contact with air was the advocated preventive measure.

In 1865, Thomas Anderson, professor of chemistry at Glasgow, drew Lister's attention to Pasteur's work on fermentation and his demonstration that living micro-organisms (microbes) and not chemical oxidation were the cause of rotting in meat. About the same time Lister learned about the use of carbolic acid (phenol) in the treatment of sewage, to render it odourless and less likely to infect cattle grazing on sewage-fertilised fields. He put two and two together. The battle against decomposition in wounds should not be fought by excluding air, but by carbolic acid dressings that could destroy micro-organisms. The new procedure was phenomenally successful, at least in Lister's hands, and one would have thought that its use would spread like wildfire through the medical profession. In fact it did not do so immediately, for many of Lister's colleagues were not equipped for the preparative precision that was needed for success — enough antiseptic to kill the bugs but not enough to burn the patient — and they often could not reproduce his results.

Truth eventually prevailed, of course. Lister returned to London in 1877 — the capital city was then still a backwater in surgical practice and needed his reforming zeal. International recognition came about the same time and

Lister received all kinds of awards. What he may have cherished most was being the guest of honour at the Jubilee of Louis Pasteur at the Sorbonne in Paris in 1892.

The Role of Empire

By 1820 the British Empire had expanded its rule to 200 million men and women, more than 25% of the total world population, and with this expansion an interest in diseases normally unknown in northern latitudes arose. Tropical medicine in particular came under the scientist's scrutiny. **Patrick Manson** (1844–1922), a Scotsman educated in Aberdeen, was one of the early pioneers who settled for some years in China studying a variety of parasitic diseases caused by the dread microfilariae, a group of microscopic worms that invade animal cells. He correctly postulated that these parasites were harboured by and developed in mosquitoes, but erroneously thought they were transmitted to humans by drinking infected water. The medical powers in London were unimpressed by Manson's studies but one convert to his concept of mosquitoes as a vector for human disease was **Ronald Ross** (1857–1932), a medical officer in the Indian Medical Service. Ross was born in India to an army family, studied at St. Bartholomew's, and returned to the sub-continent with reportedly little interest in the day-to-day practice of healing. He was more enthusiastic about studying exotic diseases like malaria, and when he chanced to meet Patrick Manson while on leave in London, the two became close associates. Ross was intrigued by the mosquito vector theory and diligently applied himself to proving that this was the means by which the malarial parasite was transmitted. To him goes the credit for identifying _Plasmodium_ (the protozoan parasite responsible for malaria), following its life cycle in the Anopheles mosquito, and recognising that mosquito bites were the source of infection. But Manson must be regarded as the driving force – he championed the theory, he encouraged and directed Ross in his work, and he finally convinced the sceptics in London. It was through his efforts that the London School of Tropical Medicine was established in 1899.

BOTANY

What could be more central to human well-being than the study of plants? They provide food, fuel, medicines. As early as 8000 B.C., Assyrians and Egyptians were expert cultivators; early Greeks developed classification schemes and wrote treatises about plants; and the Chinese became the experts in herbal medicine. In later years as Western Europe emerged from medieval influences, enthusiasm for collecting and cataloguing plants gripped scholars and lay people alike.

Recognition of Species

Religious non-conformity was prevalent in 17th century Britain, and perhaps not surprisingly many of the great scholars were of this persuasion. **John Ray** (1627–1705) was just such a man. He seemed quite comfortable and not overly ambitious as a Cambridge fellow but was stirred to action when required to take an oath he found morally repugnant. At the Restoration of the monarchy in 1660, all public figures including university fellows were required by the Act of Uniformity to swear their total allegiance to the Church of England and its teachings. Ray refused, quit his teaching job and embarked on a new career of naturalist, becoming perhaps England's greatest ever. Another Trinity Fellow, Francis Willughby, was a confederate in Ray's rebellion. Unlike Ray, he was rich and able to bestow an annual grant on his friend. The two of them formed the grandiose plan of publishing a comprehensive flora and fauna, Ray covering the plants and Willughby the animals, and they travelled together all over Europe to achieve their goal. Willughby unfortunately died young (at age 37), but Ray went on to edit and publish (under Willughby's name) treatises on birds and fishes that his friend had almost completed in addition to continuing his own previously begun catalogue of plants.

Seeing before him the uncountable *specimens* that nature presents to the acute observer, Ray recognised that they represented not a continuum, but discrete types, able to give rise, through reproduction, to individuals just like themselves — in short, he recognised the existence of _species_. He was the essential forerunner of Linnaeus, who taught us how to name the species and how to arrange them into logical groups, and of Darwin, who taught us how they evolved from one another. (Ray, of course, believed species to be "fixed", as did Linnaeus.)

Cataloguing the World

The word, _explorer_, conjures visions of Columbus discovering America and Spaniards conquering the New World. But another kind of exploration came into vogue in the British Isles where fascination with the variety of flora and fauna both locally and abroad gripped the naturalists. Explorer-botanists who travelled the world became the norm. **Joseph Banks** (1743–1820) sailed on _H.M.S. Endeavour_ with Captain Cook and brought back to Britain countless

Middleton Hall in Warwickshire. The annex building in process of restoration is probably where John Ray lived.

13

specimens from the countries they visited. But Banks's main claim to fame lies in his political influence. He recommended **Robert Brown** (1773–1858) for the position of travelling scientist to Captain Matthew Flinders who was organising a trip to Australia, and he and Brown persuaded the British Museum to house and care for their specimen collections, leading ultimately to the formation of the Natural History Museum. Banks was a leader in organising and promoting Kew Botanical Gardens and was also a close friend and associate of **James E. Smith** (1759–1828) who purchased the Linnaean collection for Britain after the latter's death. Throughout the late eighteenth and nineteenth centuries, the Kew Botanical Gardens were the major centre of plant research and classification, and two more intrepid travellers were in large measure responsible. **William Hooker** (1785–1865) and his son **Joseph Hooker** (1817–1911) were successive directors of the gardens adding their own collections of specimens from Iceland, the Antarctic, and the northern frontiers of India.

WORKING WITH NUMBERS

A concern with numbers as a means to express quantitatively the results of experiments and observations is a distinguishing feature of all modern science. But this is a field where lay people in particular find it hard to identify heroes — equations do not seem quite as exciting as observing a comet in the sky or discovering the bones of a prehistoric ancestor. One such hero, however, can be appreciated by anyone who remembers the pain of multiplication and long division before the advent of the modern calculators.

John Napier (1560–1617), eighth Laird of Merchiston in Edinburgh, was a landowner, lord of a manor with a good income from his crops and cattle, active in local and national affairs and a vigorous protagonist for the Protestants against the Papists. It was enough to keep him busy — one would think — but Napier also had a hobby, mathematics (especially as applied to numerical computation), and his hobby gained him a prominent place in history. He not only "invented" logarithms, but undertook the laborious task of computing the world's first table of logarithms, something that required more than twenty years of his life.

Napier had even earlier developed a neat device for simple multiplication, called Napier's Bones, which greatly facilitated the carrying over from one column to the next. And he played a large part in defining the columns themselves, the decimal notation we now use, with a point or comma after the integer. Decimals per se had been in use in the East for centuries before this, and they were first used in the West by the Flemish mathematician Simon Stevin in 1585, but Stevin's notation was extremely awkward — Napier's notation is the one we use today.

The generation now growing up, familiar with computers from their earliest days in school, may never be able to appreciate the indispensability of logarithms and slide rules — the latter being just logarithmic scales etched on wood or metal. They were the only calculating aids available to most scientists for more than three hundred years. There were actually two kinds of logarithms in all the tabulations, Napierian (or natural) and Briggsian. The latter were the ones we used the most, being based on the decimal system (log 10 = 1, log 100 = 2, etc.) and thus more convenient to use for multiplication and division. But the difference is a trivial matter of scale. **Henry Briggs** (1561–1630), the originator of the decimal logarithms, was professor of geometry at Gresham College in London and never claimed great originality for himself. He twice visited Edinburgh — no casual jaunt in 1625 — to discuss his proposal personally with Napier.

UNIVERSAL PHYSICS

If science were like sports, and demanded the ranking of all-time greats in an absolute order, then **Isaac Newton** (1642–1727) would be a likely choice for number one, not just in the British Isles, but worldwide. Though as a person, it must be said, he was not attractive. He was an unsociable child with few human ties and grew into a difficult and irascible adult. He never married. In the words of the biographer Richard Westfall, he was "ravished by the desire to know". He sat silently at the college table, "as isolated in his private world as though he had not come".

Newton's intellectual brilliance illuminated many fields. As a mathematician early in his career he invented the calculus, in effect creating a new mathematical language that allowed rigorous expression of dynamic concepts in science, such as velocity, force, and acceleration. In this particular historical tour-de-force, Newton was matched (in Germany) by Gottfried Leibniz, who somewhat later and independently invented a form of calculus. It is Leibniz's symbols (dy/dx) that we use today.

In the field of optics, it was Newton who discovered that sunlight is a mixture of all colours of the spectrum, which initially was not such a plausible concept even though Newton's experiments did not really permit any other explanation. He went on to express the view that light consists of small corpuscles, subject to the same laws as other bodies. This particulate view of light, however, ran into difficulties. It was opposed at the time by Christiaan Huygens in the Netherlands, who believed that light was a wave, and it was eventually "disproved" by Thomas Young in England and Auguste Fresnel in France, who showed unequivocally the wave-like properties of light rays. But then in the twentieth century Newton's light corpuscles came back into their own and the two theories were merged — both were right, and the questions that Newton himself could not answer about the behaviour of light were resolved by the new physics of quantum mechanics.

But the monumental work of Newton's lifetime was his *Principia* ("Mathematical Principles of Natural Philosophy"), one of the great enduring classics in all of science. Despite being written in Latin, it created a popular revolution, the extent of which is almost impossible to imagine today. For two thousand years philosophers and lay people alike had been convinced of the irreconcilability of heaven and earth — the natural laws in one place different from those in the other. This principle was of course buttressed by religion, but there was solid physical evidence as well — the apple falls to the ground, but the great big moon hangs in the sky forever. Newton liberated us from this dichotomy by showing that the motion of the planets (Kepler's laws) and motion on earth were governed by exactly the same principles. Central to this thesis was Newton's Universal Law of Gravitation, stating that the force of attraction between bodies of different mass is proportional to the product of the masses and varies inversely as the square of the distance between the masses. This same intrinsic force applies on earth and in the heavens, and the bodies in the sky do not collapse into one another because the gravitational force is balanced by the inertial force generated by their orbital motion.

ELECTRICITY AND MAGNETISM

William Gilbert (1544–1602) laid the foundations of magnetism as a science with his book *De magnete*, published in 1600. The phenomena of magnetism were already well known — the magnetic compass was used at sea and as early as the thirteenth century Peter Peregrinus had published a handbook, *Letter on the Magnet*. But the accepted explanation for magnetic orientation was still based on the heavenly spheres of the Greek model of the world — the magnet (called the "lodestone", the stone that leads the way) was thought to align itself with the "poles" of the outer fixed celestial spheres. Gilbert postulated that the earth was itself a giant lodestone. He demonstrated the plausibility of this view by making a small spherical permanent magnet out of ordinary magnetic material and showing that tiny magnetic needles placed on the surface of this sphere (which he named a "terrella") behave just like lodestones on the earth's surface. A particularly striking result was that the experimental needles on the terrella mimic declination, the familiar dipping of a compass out of the horizontal plane as one moves from the earth's equator towards the poles. "It has been settled by nature," Gilbert concluded, "... that in the pole itself shall be the seat, the throne as it were, of a high and splendid power."

Gilbert came from Colchester. Seen through modern eyes, he was truly an experimentalist, altogether brilliant for the time. He did not just speculate on the nature of things as was so common in his day, but actually did experiments, and the invention of the terrella has to be seen as a real stroke of genius.

Two hundred years later **Michael Faraday** (1791–1867), the shining star in the history of the Royal Institution in London and one of the genuine wonders in the history of all of science, carried out a body of experimental investigations that have perhaps no equal in their scope and brilliance and in the direct impact they had on human society. (And he was a great lecturer as well. He instituted the regular Friday Evening Discourses and the ever popular annual Children's Christmas Lectures. He himself gave the children's lectures nineteen times, at a guinea per head for adults and half that price per child.)

In chemistry Faraday discovered the rules of electrolysis and electrochemical deposition. He introduced the words *electrode, anode, cathode, electrolyte, dielectric*, and many others. In the realm of physics, Faraday used iron filings to map the lines of force associated with a magnet, and, lo and behold!, the lines were *curved*, and how do you reconcile that with Newtonian mechanics, where forces act in straight lines between two bodies? Faraday (lacking formal theoretical training) was not halted by this momentous question (as some of his contemporaries were) but instead forged ahead. He established that not only iron but many other substances respond in some way to a magnetic force. Even light was affected, he found — magnetic force alters the direction of polarisation of light.

The crowning achievement was in electromagnetism, Faraday's production of electricity from a changing magnetic field instead of vice versa. In 1820, the Danish scientist, Hans Christian Ørsted (1777–1851), demonstrated that a magnetic needle was deflected when brought close to a wire in which an electric current was flowing — the first indication that electricity and magnetism were related. Within a year Faraday reversed the experiment and what a revolution that caused! Generation of a magnetic field from an electric current was scientifically fascinating but without great practical value, but ability to generate electricity *de novo* is an entirely different matter. Electricity is our most versatile source of power and Faraday suddenly made it readily available, simply by rotating a magnet. Faraday invented electrical induction, the electrical transformer, the dynamo, the first electric motor. _All generation of electricity to this day_, whether derived from burning coal or water power or nuclear reaction, is produced by means of the dynamo and is based on the principles that Faraday recognised and demonstrated in experiments he carried out between August and December of 1831.

In the wealth of these purely experimental discoveries it is easy to miss Faraday's intellectual insight and the guidance it provided for his successors (and which, of course, guided himself to doing the right experiments). All biographers cite the fact that Faraday, by his own admission, was a mathematical illiterate. Nevertheless his mind worked _in theoretical mode_ to complement his experimental skill and everyone agrees that it was Faraday who created the idea of a physical "field of force", even if he didn't write the equations for it. As the German physicist Hermann Helmholtz put it, "Faraday performed in his brain the work of a great mathematician without using a single mathematical formula".

It was a Scotsman who took Faraday's studies of electromagnetism to a glorious conclusion. **James Clerk Maxwell** (1831–1879) is rated by present-day physicists as one of their legendary figures, comparable to Isaac Newton and Albert Einstein. Maxwell made outstanding contributions to many fields of physics, but it is the electromagnetic field that is his crowning legacy. The work was highly mathematical — essentially Faraday's rudimentary qualitative ideas put into mathematical form, and extended into a complete theory, with equations by which all interactions between electricity and magnetism could be expressed and understood. But this was not merely an _ex post facto_ unification of things already known, for it led directly to one of the momentous scientific discoveries of all time, the discovery that ordinary light (the subject of so much controversy through the preceding centuries) must be a form of electromagnetic radiation. The discovery came about because the velocity with which electromagnetic waves are propagated through space, though not directly measurable at the time, could be readily _calculated_ by means of Maxwell's equations from other factors, quantitative relations between

electric currents and the magnetic fields they generate, such as had been first measured by Ampère in 1825. The calculated result was a very large number, a velocity of 3×10^{10} cm/sec or 186000 miles per hour. More important than the number itself, the result turns out to be identical to the until then seemingly unrelated *measured* speed of light. What a revelation! "Great guns", Maxwell called it in a letter, with uncharacteristic lack of modesty.

Maxwell's work inspired a remarkable group of physicists and mathematicians who clarified and re-cast his theoretical treatment of electricity and magnetism. Among them were the Irish physicist, George FitzGerald (1851–1901), and two Englishmen of disparate backgrounds, Oliver Heaviside (1850–1925) and Oliver Lodge (1851–1940). And in Germany Heinrich Hertz (1857–1894) succeeded in generating low frequency electromagnetic waves and directly measuring their speed of propagation, confirming Maxwell's theoretical predictions and setting the stage for the later development of radio and television transmission.

HEAT AND THERMODYNAMICS

James Prescott Joule (1818–1889) came from a Manchester family that owned a brewery, which, not surprisingly, prospered with the multiplication of the city's factories and population during the industrial revolution. Both James and his elder brother Benjamin were given a good education by private tutors, including none other than John Dalton, then about seventy years old, who taught the boys natural philosophy, mathematics and some chemistry. Both brothers were financially independent for many years, so that they could afford the luxury of ambitious careers outside the world of industry or commerce. Brother Benjamin (presumably less inspired by Dalton than James) chose music and James chose science, setting up his experiments in the homes where he lived and occasionally in the family brewery.

Joule's fame rests on his establishment of the equivalence between the then entirely separate notions of mechanical energy (mechanical work done) and heat (the poorly defined "something" that could raise the temperature of matter). He measured the temperature change induced in water when mechanical work was done, which is equivalent to measuring the heat produced, and determined the ratio between the two quantities, in the distinct units then used for them. He used a paddle-wheel, turning against friction in water in a vessel with baffles designed to maximise that friction. He did the same experiment, with mercury in the vessel instead of water. In another experiment he forcibly rotated a small electromagnet in water between the poles of another magnet, working this time against the magnetic force that would by itself keep the electromagnet in fixed orientation. In other experiments he measured the heat produced by metal grinding against metal and by current flowing in wire. The experiments were done with what was for the time an awesome precision, with meticulous awareness of and correction for possible sources of error — leakage of heat from his apparatus into the surroundings and that sort of thing. His result was always the same, regardless of the manner of work being done, 772 foot-pounds of work produced one British Thermal Unit (BTU) of heat — a BTU being the amount of heat needed to raise the temperature of one pound of water by 1°F.

These were among the most extraordinary experiments ever done in classical physics. They contain the germ of the law of conservation of energy, for example, formally stated a little later by Heinrich Helmholtz in Germany, but clearly recognised by Joule in essence. He expressed it as the indestructibility and selfsufficiency of "natural powers" — only God can destroy (or create) the agents of nature.

Equally important, Joule's results settled the vexing controversy about the intrinsic nature of heat. The classical conception was that heat is a substance ("caloric"), much like a chemical element. Lavoisier had said that in 1789. Carnot had assumed it (almost thought he had proved it) in his seminal 1824 experiments on the conversion of heat into useful work. The contrary idea that heat is energy, the energy inherent in molecular motion, was recognised as plausible, but there were deemed to be no compelling arguments in its favour. Joule's results left no room for doubt. The same paddle wheel could be used over and over again, heat was produced every time that work was done, nothing emanated from the substance of the wheel or from the liquid in which it turned.

Joule was remarkably provincial, rarely leaving the city of Manchester and certainly unappreciated by his London contemporaries. He did, however, attend meetings of the British Association for the Advancement of Science which had a deliberate policy of spreading science to the provinces and generally met in places remote from London, such as York, Plymouth, Liverpool, etc. Joule read his paper on heat and work to the Association's 1847 meeting in Oxford and it proved to be a fateful occasion, for young **William Thomson** (1824–1907), later to be Lord Kelvin, was in the audience and immediately recognised the importance of what Joule had done. This encounter led directly to Thomson's subsequent preoccupation with thermodynamics.

Kelvin's great contribution to pure science is in the field of thermodynamics — he didn't invent the electric refrigerator, but he founded the branch of physics that one has to know in order to design such a device. His main achievement can be formally described by saying that he established an *absolute* temperature scale, independent of the stuff we use to measure "hotness" in ordinary thermometers. Ordinary thermometric substances all have somewhat individual responses to "hotness" and therefore tend to disagree with one another except over very narrow ranges, and the absolute scale (appropriately named the Kelvin scale) removes such ambiguity. Its importance, however, lies not in that, for ambiguity in measurement could have been removed by international agreement. What Kelvin did was to provide an entirely new concept of what temperature is — to make it an intrinsic physical property of all matter, which exists even if no-one is interested in the actual measurement of it. The definition is based on Carnot's theoretical analysis of the conversion of heat to work (as in a steam engine), which showed that useful work could be done only if heat could fall in the process from a higher to a lower temperature. The basis for Kelvin's temperature scale is this implicit potential capacity to convert heat into work. A provocative consequence of the definition is the existence of an *absolute zero* of temperature, a never quite attainable condition no matter how much effort or government dollars are expended.

INSIDE THE ATOM

A new revolution in physics took place at the turn of the 20th century — the most important event in this field since the days of Isaac Newton. Part of this revolution was theoretical and mathematical and came from the Continent; and part was experimental, emanating mainly from Britain. In Germany in 1900 Max Planck dropped the first bombshell in his Ph.D. dissertation "The Quantum Theory", the gist of which was that we must drop our preconceived notion that energy is continuously variable and imagine instead that it moves as discrete packets or _quanta_. Five years later Albert Einstein in Switzerland shook the world with his "Theory of Relativity" showing that the speed of light is beyond the conventional rules of relative motion and thereby dealing a mortal blow to conventional ideas of time and space.

In the Cavendish Laboratory at Cambridge the experimental aspect of this revolution was well underway. **J.J. Thomson** (1856–1940) led the experimental dismemberment of the atom here by his work on electrons. Many people had been studying the negative particles given off in cathode ray tubes, but these particles were generally believed to be of atomic size. Thomson actually measured their mass (strictly speaking, the ratio of mass to electric charge) and showed they were more than one thousand times smaller than the smallest known atom!

Enter **Ernest Rutherford** (1871–1937), a student of Thomson's and a native New Zealander. Under Thomson's direction he set about studying the "emanations" that came from atoms bombarded with ultraviolet light and the newly discovered x-rays. He continued these studies in his new professorship at McGill University in Canada in 1898 and later at the University of Manchester in 1907. It was at McGill and Manchester that he discovered the atomic nucleus and alpha particles (a helium nucleus carrying two positive charges), identified beta rays as electrons, detected gamma rays as x-ray-like "emanations", and most importantly predicted the existence of the then unknown neutron. With the knowledge of these particles, Rutherford proposed his planetary model for atoms, with a heavy positive nucleus at the centre and the electrons arranged in shells around it. This is where the aforementioned link with the new

Ernest Rutherford in the Cavendish Laboratory. "Talk softly", says the sign (and "don't touch the wires" it might well have added.)

theoretical physics came in, for Rutherford's model was theoretically impossible within the framework of classical physics – classically the negative electrons should have collapsed into the positive nucleus. The Danish physicist, Niels Bohr, provided the solution to this problem while working with Rutherford in Manchester using the quantum theory proposed by Max Planck in Germany. He reasoned that if electron energies (within the tiny space of an atom) can be gained or lost only in chunks, then the continuous energy loss of the moving negative electrons predicted by classical theory is impossible. Large changes (e.g. knocking the electron out altogether) can occur, but gradual spiralling is not allowed.

In 1919 Rutherford began a new phase of atomic research, the reversal of atomic disintegration. Can we bombard an atom with the normally emitted rays, he wondered, and get something different? He did indeed. He hit a nitrogen atom with an alpha particle and created oxygen-17 and a proton. Transmutation of the elements — the goal of alchemists far back in time — was realised! That same year Rutherford succeeded his old mentor, Thomson, as director of the Cavendish Laboratory in Cambridge where he continued to work until his death. His influence on modern physics extended far beyond his own discoveries through the numerous physicists of this generation whose careers began in Rutherford's laboratory — Niels Bohr, already mentioned, James Chadwick, who finally discovered the elusive but predicted neutron in 1932, John Cockcroft and E.T.S. Walton, who built the first particle accelerator, to name only a few.

ASTRONOMY

The 1639 Transit of Venus

Here we have **Jeremiah Horrocks** (1618–1641), a genius, a prodigy. Horrocks's historic achievement is his observation of the transit of Venus across the face of the sun on 24 November (old style) of 1639. Horrocks used a telescope (bought the year before for half-a-crown) which he mounted in the window to project the solar image onto a screen in his darkened room. What was so momentous about that? First of all, we have to understand that transits of Venus are extremely rare — there have been only four of them since Horrocks: in 1761, 1769, 1874 and 1882, and the next one is not due till the year 2004! Secondly, Johannes Kepler's justly celebrated *Rudolphine Tables* (published in 1627), while they predicted with an error of only two days the expected transit of 1761, did not predict the one of 1639 at all. But Horrocks, calculating planetary positions as a sort of hobby, made the prediction and thought he knew exactly when to expect the transit. He was particularly interested to measure the planet's *diameter* (relative to the sun) — he expected Kepler's estimate to be nearly ten times too large, as indeed it proved to be.

Horrocks was a curate at St. Michael's Church in Much Hoole, Lancashire (even though he had not yet reached the canonical age), and this presented some difficulties for him

"Without the sun I am silent" says the inscription on this sundial in memory of Jeremiah Horrocks in the church at Much Hoole in Lancashire.

in arranging to look for his predicted astronomical event. The day of the transit was a Sunday, and the curate was busy with church duties most of the day. He had to take Holy Communion and Evensong, occupying much of the early afternoon and leaving only a small window of time before sunset. But he managed in spite of this.

Unfortunately, Horrocks died barely a year later so we shall never know the real source of his genius or what he might have accomplished in a normal life span. His papers were left in the care of a friend, William Crabtree, a draper and amateur astronomer from Manchester, who saw to their later publication. Crabtree himself observed the 1639 transit of Venus having been forewarned by his friend of the calculated time of the event.

Astronomers Royal

The Royal Greenwich Observatory is one of the great institutions founded (in 1674) by Charles II, still flourishing though it has moved away from its old Greenwich site, which is now a part of the National Maritime Museum. Over the years this venerable institution has given us the prime meridian of the world, the zero of longitude, relative to which all others are measured, and Greenwich Mean Time (GMT), the standard by which all other times are calibrated. It has also given us (in the post of 'astronomer royal') a steady succession of distinguished astronomers. The first was **John Flamsteed** (1646–1719), who served for 44 years, the best remembered is **Edmund Halley** (1656–1742).

Halley had already had a long and productive career before he was appointed Astronomer Royal at the age of 64 — cataloguing stars of the southern hemisphere; observing a new comet and calculating its orbit, accurately predicting its return 77 years later; sailing the world to measure magnetic deviations. But it is sometimes argued that his most influential act came as a quite young man, when he proved to be the moving force behind the publication of Newton's _Principia_. It was 1684 and Halley, Hooke and Wren (at a Royal Society meeting) were discussing the problem of how to derive Kepler's laws of planetary motion, but they couldn't arrive at a consensus — Hooke, as usual, claimed he had solved the problem, but refused to reveal how. Halley was in Cambridge shortly thereafter and posed essentially that question to Newton, who, to Halley's amazement, told him without hesitation how it had to be done. "How do you know it?", Halley asked. "Why, I have calculated it," Newton replied. When Halley wanted to see the calculation, Newton was not able to find it among his papers, but promised to repeat the calculation. When he did so, he found he had made an error — the revision made for Halley was the first correct version. Halley then persuaded Newton to publish the entire *Principia* and not only served as editor but saw the manuscript through the tedious publication process.

Discovery of the Planet Uranus

In 1781 the city of Bath, a fashionable health resort without academic pretensions, witnessed the discovery of the planet Uranus, the first new planet since antiquity. The other planets known at the time (Mercury, Venus, Mars, Jupiter, Saturn) are clearly visible to the naked eye and their vagabond motions against the distant stars had been studied and recorded since time immemorial. There had been centuries of debate about their actual sizes, distances between them, etc., but the observational data had always remained the same, not significantly changed from what had been recorded by Ptolemy in Alexandria around A.D. 140. Needless to say, the discovery of a new planet — 1.8 *billion* miles from the sun, compared to the earth's 93 *million* — created a sensation and the discoverer, **William Herschel** (1738–1822) acquired great fame.

Herschel came from Germany, where he had been an oboist in the band of the Hanoverian guards. He continued to play music in England, first in Leeds and then as organist of the posh Octagon Chapel in Bath. Settled in the latter city, he found time to pursue a long-held private obsession with astronomy. He proved to be a skilled craftsman and learned to make reflecting telescopes of unequalled light gathering power, grinding and polishing his own mirrors. In 1772 he brought his sister **Caroline Herschel** (1750–1848) from

Hanover to join him and, with her aid, set himself the task of systematically scanning the sky and cataloguing special features, such as double stars and nebulae. Caroline kept track of comets.

It was during his second multi-year scan, in 1781, that Herschel encountered his new object in the sky, clearly not a star, but also not one of the known planets — what could it be? Herschel, lacking theoretical training, did not know how to calculate an orbit for his object and rather lamely decided it must be a comet; it was other astronomers (at the Royal Society in London), with better theoretical knowledge than Herschel, who decided that the object must be a planet circulating the sun. In spite of his misinterpretation, it was Herschel who seems to have gained the major share of the credit. He was also politically astute, for he named his discovery _Georgium Sidus_, in honour of fellow-Hanoverian king George III. The name, we know, didn't stick, but it did bring its reward in the form of a royal pension of £200 a year, with an extra stipend of £50 for his sister. Herschel was required to live near Windsor Castle to earn this stipend and to allow the king to look through his telescope occasionally and to explain to him what he saw.

Caroline eventually conducted her own independent research, discovered eight new comets, was the author of a revision of the standard catalogue of stars, and received numerous honours, including in 1828 — at age 78 — the gold medal of the Royal Astronomical Society. (But she did not make it into the Royal Society! It was 1945 before the first women were admitted into that august body, which is shamefully late, given the Royal Society's traditionally liberal policy in the election of male fellows.)

Space is Curved

In 1915 Albert Einstein published his "General Theory of Relativity", an event that revolutionised the way people thought about the universe. One important prediction of his theory was that a ray of light (from a star, for example) passing close to the sun would bend inwards due to the nearby gravitational field and the curvature of space. It is, of course, only possible to view such a light ray during a solar eclipse, but fortunately one was expected in May 1919 in the Southern Hemisphere. The Royal Society sent an expedition to the west coast of Africa to test Einstein's hypothesis by photographing the light from a star of precisely known position. The renowned astronomer, **Arthur Eddington** (1882–1944), was placed in charge. He later described the moment when the first measurement was made as the most memorable of his life. Einstein's theory had predicted a displacement of the light ray of 1".72 — the measured value was 1".75± .06, an extraordinary confirmation of what was then still a highly controversial theory! Since that time, the effect of gravitational fields and Einstein's theory of curved space has been tested in far more sophisticated ways, but one can imagine the pride of the British scientific community when one of their own provided the first experimental confirmation of such far-reaching ideas.

Eddington was a great populariser of science, and when he returned from the South Atlantic, he extolled the virtues of a world in which British scientists had travelled thousands of miles to prove the theory of a German colleague, at a time when the two countries were still at war. He became recognised as one of the pioneers in studies of the origin and evolution of stars, and he wrote extensively on philosophy and our place in the universe.

GEOLOGY AND PALAEONTOLOGY

During the 18th and 19th centuries the study of the earth and its history was a popular pastime in the British Isles, indulged in by amateurs as well as experts. No fancy or expensive equipment was needed; the only requirements were leisure time, an agile mind, and moderate financial resources. The usual goals were to arrange the observably distinct strata of rock in some sort of chronological order and to trace the beginning of life forms through fossils found embedded in the rock — though some went further than that and tried to understand how geological changes came about, how the successive layers were formed in the distant past.

In few other fields is the record of British work so permanently etched on the historic records — almost all the names for geological periods and epochs, now standard throughout the world, are based on names given to corresponding rock strata by British workers.

The Changing Earth

James Hutton (1726–1797) was a personification of the Scottish Enlightenment, curious about everything — not just geology, but also medicine, farming, philosophy, chemistry and economics — and ever ready to do experiments to test his ideas. Unlike many of his contemporaries and friends, Hutton held no academic appointment, but did attract devoted followers, who went with him on his geological explorations and (after his death) enthusiastically promoted his ideas. One of them was John Clerk of Eldin, a great-granduncle of James Clerk Maxwell, who accompanied Hutton on many excursions and made drawings of what they had seen. He left a portfolio of the drawings in the Clerk homestead, but they went unrecognised for what they were until 1968.

In his revolutionary *Theory of the Earth*, first made public at two meetings of the Royal Society of Edinburgh in 1785, Hutton proposed a "steady state" earth, in which geological processes act continuously at a uniform and slow rate to generate what is seen around us. Water erodes rocks, washing them into the sea, where their debris is deposited, layer after layer, and often exposed to view when the sea recedes. These deposits become heated under the enormous pressure of the weight above them and eventually melt and are thrust up in molten form through the crust of the earth, where they resolidify, ready for erosion to begin again. The sensational aspect is the apparent cyclic nature of the changes in the earth's surface and the implication that the earth must be very old. Up to that time most people had taken the biblical story of creation more or less on faith and with it an estimate of about 6000 years for the age of the earth. Hutton's proposal was not to change this by a mere factor of 10 or 100. He proposed that the earth had been here virtually for ever. His famous dictum, "no vestige of a beginning, no prospect of an end", utterly changed the way we think about ourselves and the globe we inhabit.

Hutton's key innovation was his interpretation of hard rocks (such as granite), which lack the marine fossils of sedimentary rock. Everybody thought that they must be part of the original surface of the earth, unchanged since the day of Creation, but Hutton knew that erosion indiscriminately attacks all kinds of rocks, and could not accept the implied immunity of these particular ones. Moreover, he found evidence in support of his own cyclic theory in many places, in the form of "unconformities" in geological sections, *vertical* layers of rock below the more common horizontal strata, indicative of folding and upthrust from the depths below.

Hutton's ideas were not accepted immediately, partly because he lacked the skill of easy-to-read writing. His *Theory of the Earth* became widely disseminated only after John Playfair (one of Hutton's devoted followers) wrote a clear and concise account of it in 1802, five years after Hutton's death.

A generation after Hutton, his ideas were brought to fruition by **Charles Lyell** (1797–1875). His *Principles of Geology* was first published in a four-volume edition between 1830 and 1833, and an astonishing eleven subsequent editions (some with extensive revisions) were produced during Lyell's lifetime. The reason for the book's success and the reason for Lyell's enormous influence was that *Principles of Geology* was not (in the modern fashion) an inoffensive, balanced account of existing ideas — Lyell unambiguously threw out what seemed to him to be absurd and made decisive choices between what was left. His textbook opted for what is essentially Hutton's theory of the earth. "An attempt to explain the former changes in the earth's surface by reference to causes now in operation," is how he says it in the subtitle of the book. But he pushed Hutton's doctrine to an extreme. Hutton supposed that the same kind of forces keep operating over and over again, but Lyell believed in addition that the *rate of change* had always been the same and that put him into direct conflict with the "catastrophism" theory forcefully propounded by Georges Cuvier in France. Catastrophism saw geological history as periods of little change, punctuated by natural disasters. Today we see evidence of both uniform change and catastrophe, the latter demonstrated most spectacularly in the extinction of the dinosaurs by a proposed collision of the earth with an asteroid 65 million years ago.

Lyell was an indefatigable and purposeful traveller, continually subjecting his ideas to test. To call it experimental testing would be perhaps too strong a term, but he could not or would not believe any assertion of fact or inference without going to see for himself, questioning the original proponent and anyone else who might have been an informed bystander. He went on long arduous trips, virtually every year of his life, no matter what other duties he might have. Wherever he went he sought out scientists of whom he knew, but he sought out knowledgeable people even if there were no affirmed experts. His friend Roderick Murchison had given him good advice — every village includes a naturalist, he had told him, and to find him, start by asking at the chemist's shop.

The Fossil Record

The fossil record is essential to any attempt to describe successive layers in the rocks and one of the earliest scholars to realise this was **William Smith** (1769–1839), sometimes called the father of English geology. Smith was a surveyor and civil engineer who had to work for a living and certainly was not endowed with much leisure time. However, he did manage to combine his surveying work with an interest in geological rock formations and particularly with the fossils occurring in the various geological strata. He was an exact contemporary of the more famous Georges Cuvier in France, and they both, independently, were responsible for the recognition that the fossil record corresponded with stratigraphical successions thus providing an independent means of dating geological formations. It was not easy for Smith to break into the then "geological brotherhood" and it was only late in his life that he was recognised by the London Geological Society through the award of their coveted Wollaston Medal in 1831.

The fossil record was an important factor in Lyell's thinking and he interpreted it as providing a firm and convincing foundation for slow and gradual change. Lyell (in collaboration with the French palaeontologist Paul Deshayes) made a major original contribution by statistical comparison of mollusc fossils as a function of where they were found. There was a high proportion of still living species near the top of the present surface of the earth's crust, but a progressive increase in extinct species in lower deposits. On the basis of these results Lyell subdivided the Tertiary era in geological stratigraphy (the period beginning about 60 million years ago — see Table 1) into the still standard subdivisions of Eocene, Miocene, Pliocene and Pleistocene periods. (The root "cene", incidentally, means "recent" — the entire tertiary era is recent on the geological time scale.)

Table 1 Geological eras

Cenozoic Era	
Quaternary	up to the present
Tertiary	less than 65 million years ago
Mesozoic Era	
Cretaceous	65 million years ago
Jurassic	136 million years ago
Triassic	190 million years ago
Palaeozoic Era	
Permian	225 million years ago
Carboniferous	280 million years ago
Devonian (Old Red Sandstone)	345 million years ago
Silurian	405 million years ago
Ordovician	425 million years ago
Cambrian	550 million years ago

During the 19th century frenetic activity in earth history took place throughout the British Isles, particularly in studying rock formations from the Palaeozoic era (600–200 million years ago). **Hugh Miller** (1802–1856), a Scottish journalist, explored what was called Old Red Sandstone (formed approximately 350 million years ago) throughout the highlands of Scotland. In the south, in Wales and the border regions, **Roderick Murchison** (1792–1871) and **Adam Sedgwick** (1785–1873) described, respectively, the Silurian and Cambrian periods which formed when Wales and the border counties had been covered with a shallow sea. Murchison was a retired army officer of independent means who came late in his life to the study of geology. He met Adam Sedgwick, the Cambridge professor of geology, through their membership in the Geological Society and was more or less taken under Sedgwick's wing in the early days of their association. Together they set out to explore the old rock regions in Wales and the border counties but soon went their separate ways, Sedgwick working in the north west of Wales and Murchison in the south and border regions.

Scientists, even then, were contentious individuals concerned with priority of discovery and challenging each other over interpretations of their observations, so, perhaps predictably, it was not long before Murchison and Sedgwick fell out over the issue of which had found the earliest life forms. Surprisingly, they later found themselves on the same side of an even more ferocious argument regarding the age of the Devonian strata studied by **Thomas De la Beche** (1796–1855). The latter claimed that his Devonian strata, which contained vegetable matter, was older than Murchison's Silurian strata which did not. Why should this be important? Aside from the purely intellectual pursuit of correctly tracing earth history, there was a practical matter — coal. If carboniferous material were present in strata older than the Silurian, it suggested coal should be sought in strata previously thought to have none. Eventually De la Beche was shown to be wrong in his dating and the matter was settled, but not without a number of acrimonious years within the elite membership of the Geological Society.

Dinosaurs and Relics of the Flood

There were many motivations for fossil collecting, some quite unrelated to stratification. **William Buckland** (1784–1856), Oxford's most famous (or should one say notorious?) professor of geology, searched far and wide for what he considered relics of the Deluge — his popular _Reliquiae Diluvianae_ was published in 1823, subtitled "Observations on the organic remains contained in caves, fissures, and diluvial gravel, and on other geological phenomena, attesting the action of an universal deluge." Ironically, Charles

Lyell, who swept away all notions of diluvianism in the 1830s, had been an enthusiastic student of Buckland's from 1816 to 1818.

Of greater long-range significance was the discovery about this time of giant skeletons, attesting to the former presence on earth of animals of incredible size and bizarre shapes — the dinosaurs ("fearsome lizards") and their marine analogues, the ichthyosaurs. The first such skeleton (of the marine variety) was found in 1812 by a young girl, Mary Anning, in Lyme Regis. She was soon adopted by William Buckland and the two of them became a common sight, wading in search of fossils in the shallow waters off the rocky Lyme Regis shore. Descriptions of their finds were published by **William Conybeare** (1787–1857), a distinguished clergyman and geologist who drew reconstructed pictures of whole animals on the basis of the ichthyosaurus and plesiosaurus fossils and called attention to their similarity to the living crocodiles. The first proper description of one of the spectacular terrestrial dinosaurs was given by **Gideon Mantell** (1790–1852), based on remains he found in Tilgate Forest, about half way between London and Brighton.

Dinosaur fossils were soon discovered in all parts of the world, attesting to widespread distribution, and they were never associated with human remains, indicative of an early period in terrestrial history when humans did *not* exist, but giant creatures, *now extinct*, dominated the whole globe. The question of whether these findings could be made compatible (by those who wished it) with the biblical story of creation is not really important. The important point is that the dinosaurs gripped the imagination of the British public as no other scientific discovery had ever done. Models were built for the great Crystal Palace Exhibition of 1851, books about them were avidly read. The influence on the popular awareness of biological science must have been enormous — surely this prepared the ground for the later ready acceptance of evolution and all that.

EVOLUTION AND ANTHROPOLOGY

The Origin of Species

It has been said that Darwin was to the 19th century what Galileo and Newton had been to the 17th — creator of a previously unimaginable revolution in our conception of the natural world. The statement itself is surely indisputable. Darwin was as devastating a force in history as any military man.

However, in deference to historical accuracy with respect to the scientific ideas per se (as opposed, shall we say, to *popular* influence), it is not quite fair to use the name Darwin by itself, without the acknowledgment that Wallace had the same idea independently at about the same time. And both men received their inspiration from the same source — overseas journeys away from England, in this case far beyond the familiar European scene to virtually unexplored terrain in South America and/or Malaysia.

Charles Darwin (1809–1882), the man we have just called a "devastating force" in history, was by no means a conquering hero — he sounded no trumpets, he beat no drums. He was a lifelong hypochondriac, an indecisive ditherer. It would be wrong to label him as reluctant to *reveal* his theory of the origin of species, for he wrote it all down in a 230-page manuscript in 1844, and described and discussed it freely in his correspondence. But formal publication was another matter. Imagine yourself in his position, having just upset the applecart of conventional opinion about creation and the smug certainty of man's position at its apex. What would you have done? What Darwin did was to devote the following eight years of his working life to the taxonomy of barnacles! It was not until June 1858, when the bombshell of Alfred Wallace's paper with the same theory arrived by post at Down House, that Darwin finally got down to the task of publication — the story of that and the persuasive role of Darwin's friends, Charles Lyell and Joseph Hooker, is too well known to require repetition here.

Darwin's faltering sense of resolution is evident from his earliest days. His father, a successful physician, sent him to Edinburgh to study medicine, but that enterprise failed and he was shipped off to Cambridge to become a clergyman instead. Darwin began his beetle collection there, but had little enthusiasm for the prescribed studies — by his own admission, he went to very few lectures and had little interest in most subjects. He did form a decisive friendship in Cambridge with the botanist John Stevens Henslow, and it was the latter who recommended him for the position of unpaid naturalist on the Royal Navy's surveying ship, *H.M.S. Beagle*. Darwin was 22 years old when the five-year journey began, regarded himself (correctly) as better prepared to survey geology than flora and fauna, and took along the just published first volume of Lyell's *Principles of Geology* as reading matter. His geological observations and a new hypothesis about the origin of oceanic islands and coral reefs were the first things he published after his return to England in 1836. Darwin was secretary of the Geological Society in London from 1838 to 1841; a book, *Structure and Distribution of Coral Reefs*, came out in 1842.

Darwin, now ensconced at Down House, with no further thought of travel, next turned to the acute observations of birds and other animals that he had made on his voyage and to the theory which had gradually crystallised in his mind. Evolution itself was nothing new, contrary to much present popular opinion. Darwin himself, in an introduction to *The Origin of Species*, lists 24 scientists who had proposed evolutionary ideas, beginning with Lamarck in 1809 and even including Richard Owen, who was soon to become Darwin's bitter opponent in the arena of public acceptance. The uniqueness in Darwin's proposal was a plausible *natural mechanism* for evolution, i.e., natural selection, survival of the fittest.

Origin of Species, when finally published, became an immediate sensation, and was ultimately translated into 29 languages. Not everyone was enthusiastic, of course, and we all know that there are still vociferous detractors today. Politicians at the time must have had not a little discomfort, for Darwin was never offered the knighthood he had every right to expect. (Three of his ten children were knighted after his death.)

Alfred Russel Wallace (1823–1913) had the idea of natural selection *independently* of Darwin. But he was a much younger man and lacked Darwin's status — no illustrious grandparents, no university degree — and he is perhaps given less than his fair share of the credit for the discovery. On the other hand, there is no doubt that Darwin had the concept in his mind (even though it was unpublished) several years before Wallace set out on his first journey of exploration.

Wallace was the eighth of nine children of a West Country family beset by constant economic woes. His urge for travel to remote places was first aroused by reading the works of Alexander von Humboldt and others in the 1840s — and that included Darwin's journal of the _Beagle_ voyage. He and a friend (Henry Walter Bates) decided to take off on a bold trip of their own to the Amazon basin, without sponsors or financial support, hoping to make money from the sale of exotic specimens at home. Wallace unfortunately lost most of his materials (and almost his life) when his ship sank on the return voyage, but he managed to publish a book about his trip, and, undaunted by the previous disaster, set off on another journey, this time to the Malay archipelago. Evolution of species had been on his mind right from the start of his first journey and he wrote his first paper on his hypothesis about how it worked while still at work in Malaysia — it was published in 1855, without apparently attracting much attention. The "bombshell" paper mailed to Darwin from Malaysia in 1858 was his second (and clearly more fruitful) account.

Wallace, despite his many years of experience in collecting and classifying, was only 35 years old in 1858, fifteen years younger than Darwin. He went on to live to the ripe old age of 90 and became a popular and respected figure in the world of natural history — despite some adventurous excursions into spiritualism, socialism, and the like, which diminished his reputation in some quarters.

Richard Owen (1804–1892) and **Thomas Huxley** (1825–1895): We cannot leave the subject of evolution without mention of these two prominent Victorian naturalists, whose bitter warfare pro and con helped enormously to sharpen public awareness of the subject.

Richard Owen, lacking private wealth, had to work for a living and did so as a museum man. He became supervisor of the natural history collections at the British Museum and in this job acquired an enviable national reputation. He was a proponent of evolution of a sort, but apparently not Darwin's kind, for, after publication of *Origin of Species*, he became Darwin's bitter enemy — purely out of jealousy is what Darwin himself believed.

Huxley, on the other hand, was another upper class figure. He had been in the Navy and he too went on his voyage of exploration (on *H.M.S. Rattlesnake*), but no notable discoveries resulted. Huxley's fame comes in fact from his tireless efforts as writer, public lecturer, and, above all, crusader for Darwin and his message. He wrote the obituary notice for Darwin in the magazine *Nature*, expressing a judgement that in fact he himself helped to bring about: "He found a great truth, trodden under foot, reviled by bigots, and ridiculed by all the world; he lived long enough to see it... established in science, inseparably incorporated with the common thoughts of men."

The Search for Human Fossils

The theory of Darwin and Wallace redirected many palaeontologists into evolutionary channels, searching for early humanoid remains. Important discoveries were made on the Continent and in the Far East and well into the present day in Africa. It was at first rather humiliating for people to think that they were descended directly from the beasts. Honour was deemed to be preserved by the dogma that brain size created a sharp demarcation, but fossils to suggest the stepwise evolution of a large brain were lacking, until (in 1912) a skull and jawbone discovered on Piltdown Common in Sussex obligingly filled the gap, providing an ape-like jaw associated with a huge (almost human) cranial cavity, in an early Pleistocene environment, probably close to a million years old. Most of the experts accepted it at face value, vying with each other in reconstructing models of the whole body of this "missing link".

The unravelling of the plot took several decades, initially spurred by the discovery of other skulls of equal antiquity all over the world, none of which suggested that brain size had led the way in the evolution of early man — the Piltdown skull simply didn't fit. It eventually turned out to be a modern human skull, only two thousand years old, and the jawbone was practically yesterday's, subsequently identified by anatomists as coming from an orang-outang. All the bones had been carefully stained to give the superficial appearance of great antiquity.

Who was the culprit? Nobody knows for certain — many theories have been proposed, and almost everyone concerned has been implicated by one or another of them. The most recent book (1990) points the finger of guilt at Sir Arthur Keith (one of the reconstructors), who would have had ready access to both orang-outang and human remains and is said to have been driven by insane ambition for a leading place in the world of science. (If true, is there a parallel with modern fraud — fame rather than fortune as the underlying motive?)

CHEMISTRY

The Beginnings of a Modern Science

Robert Boyle (1626–1691), the "sceptical chymist", was the youngest son of the first Earl of Cork, an Anglo-Irishman who acquired his wealth and lands as an Elizabethan immigrant. He sent Robert to Eton College for his basic education and then (at enormous expense) to Geneva and Italy with a private tutor. Robert was in Florence, studying the works of Galileo, when disaster struck back home — the Irish rebellion of 1641 devastated the Earl's lands and severely depleted the family fortunes. But the rebellion was eventually put down and enough of the family fortune was salvaged to keep Robert more or less independent. He never worked at a paying job nor held any university appointment and was even able to hire private assistants for his experiments — Robert Hooke (1635–1702) being the most famous — and secretaries for his writing. He never married, which presumably helped to make ends meet.

Boyle invented (with Robert Hooke's help) the modern air pump and experimented with gases (giving us what we still call "Boyle's Law"). He was a leading dissector of animals. He introduced the use of alcohol as a preservative. He was a key member of the Oxford group that became the Royal Society of London.

Most historians believe that Boyle's most important contribution to science was his espousal of a "mechanical philosophy", especially with reference to chemistry. Chemistry was an art before Boyle's time, even an *occult* art, completely outside the framework of the world of physics, and that's what Boyle changed — he insisted that chemistry and physics were part and parcel of a single natural philosophy. His famous book, the *Sceptical Chymist*, expresses firm conviction for a particulate theory of matter — in his own words, all matter seems to be "divided into little particles of several sizes and shapes variously moved" — but the book as a whole probably made its impact more by asking the right questions than by providing answers, and by an appeal to quantitative experiment for support or derogation of theoretical ideas. The magic recipes of former chemists, lacking a theoretical basis, were dismissed as unproductive.

Nature's Constituents

Joseph Priestley (1733–1804) was an arch-dissenter, who obstinately insisted, or so it seems today, on taking the unpopular side in any argument. He is best known for his "discovery" of oxygen — which he stubbornly persisted in *refusing* to recognise for what it was. He was equally well known in his time for his provocative views on religion — his denial of the Holy Trinity was much stronger than mere "dissent" and was repugnant to many nonconformists as well as conservatives.

Priestley was the son of a cloth dresser, born in a small town near Leeds. He was educated as a minister at a dissenting academy at Daventry, and for six years he taught literature and language at another such school in Warrington. His most productive period was in the service of the politically ambitious Earl of Shelburne, formally as librarian, but actually as an aide of broader usefulness, a sort of resident intellectual. It was during this period that Priestley did most of his chemical work — he chemically defined not only oxygen, but also ammonia, hydrogen chloride, hydrogen sulphide, and several other new gases.

Priestley eventually left the Earl's service and returned to the ministry in Birmingham. There his theological views became increasingly radical and in 1791 he even expressed support for the French Revolution (implicitly favouring an end to monarchy in Britain, too?). With this he provoked mob violence and his personal safety was threatened. On the advice of his friends he moved to the now independent United States of America. His polemic writing continued from there and one of his detractors commented that "the Government of Heaven itself, should he ever get there, will, in his opinion, want reformation."

From the opposite end of the social spectrum we have **Henry Cavendish** (1731–1810), a member of the immensely rich family of the Dukes of Devonshire, who divorced himself from that family's normal public life to become an ascetic loner in the quest for scientific knowledge. He regularly attended meetings of the Royal Society and of the dining club composed of some of its members, but was almost a recluse otherwise, taking great pains to avoid most kinds of human contact. Like Priestley, Cavendish was chiefly a gas chemist, who in 1766 discovered "combustible air" (hydrogen) and demonstrated that it could be combined with a portion of "ordinary air" to form water, an event of equal importance with Priestley's discovery of oxygen.

Atoms and Elements

John Dalton (1766–1844) can rightly be dubbed the "second founder" of the atomic theory; Democritus, way back in ancient Greece, being the first. Dalton was born in the tiny and remote village of Eaglesfield in Cumbria, the son of a poor cotton weaver. William Wordsworth, the poet, was born (in 1770) in the larger town of Cockermouth, just two miles away. Poet and scientist both knew the blind philosopher and local sage, John Gough, who lived in nearby Kendal. Wordsworth wrote about him in his poem "The Excursion": "Methinks I see him now, his eyeballs roll'd beneath his ample brow." For Dalton he was a source of educational guidance and was responsible (in 1793) for getting him a teaching job in Manchester.

Dalton became interested in what one might call systematics of science before he came to Manchester and in that city found kindred spirits to encourage his pursuits. He never married and for more than a quarter of a century shared a humble dwelling in George Street with a friend and his family. The Manchester Literary and Philosophical Society was the centre of his scientific world. He contributed 116 papers to its proceedings and was its president from 1819 until his death. He was invited to lecture at the then newly founded Royal Institution in London in 1804 and again in 1809 and made other occasional trips to the capital; he once made a short visit to Paris; once a year he took a holiday in the Lake District; but most of the time he was in Manchester, gradually becoming one of its most respected citizens.

Dalton's great classic, *A New System of Chemical Philosophy*, was published in several volumes, the first appearing in 1808. The atomic theory of chemical combination was only a part of the book, but it was soon recognised as the vital part. Atoms are seen as tiny (invisible) spherical bodies of fixed mass; each different *chemical element* has its own distinct kind of atom; atoms combine in *definite proportions* to form molecules (which Dalton called "compound atoms"). Dalton's measurements of relative combining weights were crude and his assumption that simple molecules have a 1:1 stoichiometry (water = HO) distorted his figures for relative atomic masses, but those are trivial faults. What counts is Dalton's conceptual revolution — he established the model which chemists have used ever since to visualise and think about what goes on in their reaction flasks. Derek Gjertsen quotes a 19th century student's answer to an examination question: "Atoms are blocks of wood, painted in various colours, invented by Dr. Dalton." The only difference today is that the blocks are made of plastic.

Humphry Davy (1778–1829) came from Cornwall, from a family of unpretentious yeoman stock. He was largely self-educated and was fortunate as a youth to find science-related employment at Thomas Beddoes' Pneumatic Institution in Clifton (close to Bristol) — an institution dedicated to treating diseases by the inhalation of various gases. It was there that Davy discovered the anaesthetic properties of nitrous oxide ("laughing gas"). It did not become (as he had hoped it might) a widely used medical agent, but it did bring him fame by a different and unexpected route — sniffing the gas became a popular rage.

Davy's most inspired work was carried out while director of the Royal Institution and was a brilliant application of Alessandro Volta's voltaic pile, discovered in Italy a few years earlier. The continuous source of electricity which it provided could be used to disrupt chemical compounds (e.g., H_2O into oxygen and hydrogen) and Davy enthusiastically poured his energies into exploration of the new possibilities thereby created. He found that he could use the pile to decompose the previously intractable "earths" to produce shiny, quicksilver-like globules of a whole range of new metals — sodium from soda, potassium from potash, calcium from lime, and several others. It was a discovery of the first magnitude, deliberately sought, based on a kind of comprehension of what chemistry was all about that would have been unthinkable even a decade earlier. Davy, when he first saw the metal break through the earthy crust, is reported to have danced about the room in ecstasy. Napoleon gave Davy a medal for the work and Davy went to France to collect it in person, despite the fact that England and France were engaged in a bitter war. Davy was subjected to some criticism for his breach of the battle lines but swept it aside: "The two countries or governments are at war," he said, "the men of science are not" — a noble sentiment which today's world might not be inclined to accept with tolerance.

From the late 18th century onwards new chemical elements were discovered by scientists all over the world. Sweden holds the record for the largest number but Britain was not far behind. **Smithson Tennant** (1761–1815) identified iridium and osmium during the course of purifying platinum to build boilers for sulphuric acid production, and his friend and collaborator, **William Wollaston** (1766–1828) contributed palladium and rhodium to the growing number of new elements. **Norman Lockyer** (1836–1920) and **Edward Frankland** (1825–1899) found a new element during spectroscopic studies of the sun and appropriately named it helium. The Scottish chemist **William Ramsay**

(1852–1916) discovered argon, krypton, neon and xenon. And, somewhat later, **F.W. Aston** (1877–1945) was the first to identify isotopes of non-radioactive elements.

Order from Chaos

Chemists were consumed not only by the attempts to discover new elements but also by the need to make some sort of sense of this array. Priority for success in this latter area should probably go to **J.A. Newlands** (1837–1898) who constructed the first "periodical table" displaying atoms in *increasing order of atomic weights* and noting what was called "the law of octaves" — i.e. element eight in any sequence had similar properties to element one, and this similarity extended to every seventh element thereafter. This idea was scorned by most of his contemporaries, one physics professor asking whether Newlands had ever examined the elements in *alphabetical* order to see if he could find any regularity that way. Similar work by Mendeleev in Russia nearly five years later finally convinced the sceptics.

The organisation of atoms in chemical compounds was another thorny problem that concerned chemists of the day, and this too was solved by an Englishman — Edward Frankland, mentioned in connection with the discovery of helium. He was the first to recognise that atoms of chemical elements can only combine with fixed numbers of other atoms — the beginning of the concept of valency and a landmark step on the way to understanding the nature of chemical bonds.

ENGINEERING AND TECHNOLOGY

A profound change in British life took place in the last half of the 18th century as the country moved toward industrialisation. Coal, iron, and textiles were the bases of new sources of wealth together with the associated need for transport infrastructure. And transport did not mean only roads and railways; water was important, too, as evidenced by the Duke of Bridgewater's ship canal linking Manchester and Liverpool.

Many of the people who played leading roles in the explosion of technological developments were not scholars in the usual sense, some were not even formally educated, and their motivation was not discovery of the laws of nature, but the practical considerations of money and fame. As early as 1705 **Thomas Newcomen** (1663–1729), an iron-monger from Dartmouth, built a steam engine that operated at atmospheric pressure and, though not very efficient, was used to pump water from the mines in the South West. Best known to the layman, perhaps, are those pioneers who improved his design, **James Watt** (1736–1819), **Richard Trevithick** (1771–1833) and **George Stephenson** (1781–1848). Watt was a Scottish born instrument maker who left his position as "mathematical instrument maker" at Glasgow College to go into partnership, first with Dr. John Roebuck of Carron and later when Roebuck went bankrupt with Matthew Boulton in Birmingham. His ideas for improving the efficiency of Newcomen's steam engine involved among other things insulating the steam cylinder to prevent heat loss and introducing a partial vacuum in the steam condenser. Watt's machines, like those of Newcomen, were used primarily for pumping water from mines, but it was not long before both Trevithick and Stephenson turned to designing steam engines for moving people and goods from one place to another. Trevithick, the son of a Cornish mine manager, has been described as a slow and obstinate scholar, but he had the creative spark that led him to realise that greater efficiency could be obtained by using high-pressure steam rather than steam at atmospheric pressure. He has the distinction of devising the first steam-drawn carriage on a railway. Stephenson was a Newcastle lad who didn't learn to read until he was 18 years old, but he effectively made trains the bedrock of the transportation system essential to the growing manufacturing economy.

Although none of these inventors were pure scientists in the usual sense, there is an interesting connection to pure science. Joule and Thomson and their French and German colleagues in the formulation of the laws of thermodynamics (*see* HEAT & THERMODYNAMICS, above) were influenced as much by the performance characteristics of steam engines as by purely natural phenomena. Interplay between basic and applied science is a two-way street.

Massive construction projects continued throughout the 19th century with **Isambard Kingdom Brunel** (1806–1859) at the forefront building bridges, piers, docks, steamships, and, of course, the Great Western Railway. Brunel, unlike many engineers of his day was the son of a famous inventor, Sir Marc Isambard Brunel, and he was not self-taught but was sent to Paris to learn his trade.

Chemical Industry

Building and engineering advances were not the only things that marked this period of change. Chemistry came into its own as an applied science with the development of the alkali industry for bleaching textiles, the erection of massive soap factories around the city of Liverpool, and the synthetic production of noxious materials like sulphuric acid.

Even academics became involved in chemical technology. With remarkable foresight the government established a practical school of chemistry in 1845 and lured the distinguished German chemist, August Hofmann, to London as its director. Many of his students were to become leaders in chemical industry. **William Perkin** (1838–1907), in particular, set off a flurry of activity when he accidentally *synthesised* a new textile dye — raising visions of huge profits for factories as natural dyes were eclipsed. However, the dye industry (both natural and synthetic) did not last long in Britain. It was cheaper to export coal-tar raw materials to Germany and import the dyes made from them than to carry out the entire process itself. Later on, as German methodology advanced, home manufacture of synthetic dyes became prohibitively expensive because of huge patent fees demanded by German chemists.

August Hofmann returned to his native country abandoning the Royal College of Chemistry not long after Perkin's synthetic success. He deplored the apathy of the government towards chemical research and the inability to see that commercial success rested upon the creativity of scientists. In the ensuing years Germany, not Britain, led the way, not only in chemical dyestuffs, but also in many other areas of chemical technology. The folly of putting short-term profits before long-term prosperity is not a new phenomenon!

MOLECULAR BIOLOGY

The Birth of Molecular Biology

The elucidation by Watson and Crick in 1953 of the molecular structure of DNA, the celebrated double helix, was surely the greatest biological discovery of the 20th century. Curiously, it happened in Cambridge's Cavendish Laboratory, the very same place where so much history was made in pure physics.

The DNA problem goes to the core of the basis of life. The genetic information carried by the DNA of a single cell contains billions of bits of information — the blueprint for how an organism is built and how it functions. It only takes a difference of about 1 bit per 1000 to mark one individual

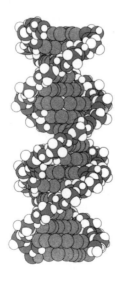

The DNA double helix — the secret of inheritance.

human being as distinct from another; differences of the order of 1/100 take us from one animal species to another; differences of the order of 1/10 from an animal to a plant. A single mutation (only one bit altered) at a critical spot can be lethal. How does DNA code this information? How are true error-free copies made over and over again when cells divide? How is the paternal and maternal mix of characteristics achieved in sexual reproduction? How is the information within DNA expressed? (We now know that proteins, not DNA, are the *working molecules* of living cells, which determine what a cell is capable of doing. The last question in the foregoing list can thus be paraphrased: How can DNA dictate the primary structures of proteins that a cell actually synthesises?)

The bold step taken in answering these questions — the genius, if you like, of Watson and Crick and the underlying Cambridge philosophy — was to go directly to the three-dimensional atomic structure of DNA, bypassing all kinds of unsolved problems that still lined the road between genetics and chemistry. Watson and Crick did not themselves do the experimental x-ray diffraction studies which formed the basis of the structure determination, but they did the interpreting of the photographic array of spots — did it with their minds and pencil and paper, for the computer programs one would now use had not yet been written. The double helix was the result, and, as it happens, this particular structure by itself virtually solves the problem of how genetic material replicates itself — the mutual alignment of the two helices leaves very little to the imagination.

Once this problem was solved, a whole army of scientists all over the world were able to work out how DNA code is translated into the structure of proteins, how DNA is damaged and repaired, and how it can be mutated, all the processes necessary for survival of living organisms. This is a true example of the old adage "a picture is worth a thousand words" — the helical structure allows scientists to ask questions in an entirely new way and to design problem-solving experiments that they could never have thought of without the picture before them.

Viruses, Muscles, and Nerves

Cambridge continues to stand in the forefront of research in molecular biology. Atomic arrangements within the molecules of viruses are gradually being elucidated. The molecules involved in muscle contraction are being identified and studied. The processes of nerve stimulation and conduction are gradually becoming understood. "Structure" is still the key word and x-rays are still a primary experimental tool. But no other biological molecule has ever been found to yield its secrets as readily as DNA did — the simplicity of the double helical structure was a unique result.

Part II
Places to Visit

1
London

Entries for London are grouped as follows:

1. The "City" — Tower of London and St. Paul's
2. Inns of Court to Westminster
3. Piccadilly and vicinity
4. British Museum and Gower Street
5. Kensington Museums to Paddington
6. Outer areas — in alphabetical order

Sites within groups (except group 6) are arranged in convenient sequence for a walkabout.

1. The "City" — Tower of London and St. Paul's

This section covers most of the true "City of London" and much of it bears the scars of the great fire of 1666. The area includes the first seat of the Royal Society and the sites of several trailblazing discoveries. We begin at the Tower of London and move northwest to Holborn and the Charterhouse, a total distance of about 1.5 miles.

ROYAL MINT (Tower of London)

When Isaac Newton was appointed Warden (later Master) of the Mint, it was situated in the Tower of London — in the cramped space between the outer wall and the inner bailey — and there was a residence for the Warden right next door. Newton lived there for several months, but he found the noise and dismal surroundings too much for him and soon moved to healthier quarters in the Piccadilly area. The Mint itself moved to a new location outside the walls in 1810 and in 1967 departed London altogether; its present headquarters are at Llantrinsant, near Cardiff. (An interesting historical note is that the old Mint was considered a place of sanctuary in England and it was not until after Isaac Newton took office that laws were passed to abolish this status.)

The inter-wall area is not open to the public but is easily seen from near the beginning of the customary tour through the Tower. Visitors usually enter the Tower precincts at the southwest corner and proceed through an arch at Byward Tower into the walled area towards the Bloody Tower. Just beyond the Byward Tower arch a narrow street is seen on the left, marked "private" because it is part of the residential area for yeoman warders ("beefeaters"). This street is in fact the narrow space between the inner and outer walls where the Mint was located — on maps it carries the name of Mint Street, but it is not marked as such. The Mint itself began at Legge's Mount, the farthest point you can see, where the Tower walls make a sharp right turn; other buildings and the Warden's house were just around the corner. It is a pretty grim prospect even today, living on a narrow lane, facing the 40-foot-high inner wall. We can imagine how much worse it would have been when they were grinding out the nation's coinage next door!

ST. OLAV'S CHURCH (Tower Hill)

William Turner (1508–1568), the "father of English botany", was born in Northumberland; he was an outspoken religious reformer and forced into exile during the reign of

Queen Mary and for one other spell. He turned necessity to advantage during these periods by becoming well acquainted with European naturalists and their research, and was inspired thereby to produce a list of English flora and fauna, based largely on his own observations — designed to replace the derivative (often ancient) works that were then being used to teach students at the universities. When in England, Turner lived most of his adult life either at Wells (where he was Dean of the Cathedral) or in a house on Crutched Friars, just north of the Tower of London. He died here and was buried at St. Olav's church, one of the few churches in the city to escape the great fire of 1666, but badly damaged by bombing in World War II; what we see today is a sensitive post-war restoration. Turner's tomb is in the wall in the southeast corner of the church, with a noble inscription, all in Latin — though his own most famous work, the _New Herball_, was, contrary to tradition, written in English.

Curiously, the same wall carries a monument to a modern amateur botanist, a local banker named Job Edward Lously, who studied plants that started to grow in the rubble after the bombing of the 1940s and unexpectedly found several rare specimens. A more famous amateur buried here is Samuel Pepys (1633–1703), president of the Royal Society from 1684 to 1686, chiefly remembered today for his diaries, which include a description of the Great Fire and other events of the time. He lived on Seething Lane, next to the church, and worked for the Navy in a building that stood across the street. He and his wife were buried beneath the high altar of St. Olav's and both have impressive memorials.

The church is at the corner of Hart Street (an extension of Crutched Friars) and Seething Lane. It is well signposted from Tower Hill Underground station. Number 42 Crutched Friars is a substantial building surviving from the early 18th century and indicating that this was a very prosperous area at the time.

The Monument. The Great Fire started a few hundred yards west of St. Olav's, close to the site of this monument designed by Christopher Wren. The fire burned for 3 days, 436 acres were devastated, 13 000 houses destroyed. It caused the Royal Society to move away from Gresham College for 8 years into temporary quarters; it destroyed all the priceless books and instruments that had been left to the Royal College of Physicians by William Gilbert and William Harvey.

Royal Exchange.   The Exchange, at the financial hub of the city (intersection of Cornhill and Threadneedle Street), serves to remind us that Thomas Gresham, that great bene-factor of science (see p.59 below), was first and foremost a money man, proponent of the law that "bad money drives out good". It was he who created the Royal Exchange in the reign of Elizabeth I and it has remained at the same site ever since. (But the present building is a Victorian restoration.)

St. Paul's Cathedral.   St. Paul's is a monument to the genius of architect Christopher Wren (1632–1723), the bulk of whose work was stimulated by the need to rebuild London after the Great Fire. Were it not for that disaster, he might well be remembered today as a scientist, for he was in fact one of the close-knit group who founded the Royal Society (*see* OXFORD p.157) and held positions as professor of astronomy at Gresham College and at Oxford — appoint-ments made in 1657 and 1661, respectively. Appropriately, the Greenwich Observatory is one of his notable architec-tural successes.

Royal College of Physicians.   The College was founded by royal charter in 1518 (about 300 years *before* the Royal College of Surgeons), largely through the influence and actions of Thomas Linacre, who provided the first quarters for it in his own home, in Knightrider Street, just south of St. Paul's. The initial function of the College was only to license physicians (and thereby prevent quackery) in the London area, but its influence on teaching and medical prac-tice grew beyond anything the founders could have imag-ined — it played an inestimable role in the emancipation of science and medicine from religious and social intolerance. Subsequent locations of the College were at Amen corner (Warwick Lane), where the canon's houses of St. Paul's now stand, and then (1674–1825) in a grander building further down Warwick Lane, next door to the present Cutler's Hall. A plaque marks the latter site.

Most of the old Knightrider Street has been absorbed into a British Telecom building; only a tiny piece remains, between Sermon Lane and Goodman Street. The present location of the College is on St. Andrew's Place at the southeast corner of Regent's Park.

St. Bartholomew's Hospital. This historic hospital stands just a few steps north of Warwick Lane; it is the oldest charitable institution in London that retains its original site, though none of the original buildings are preserved. The peerless William Harvey (*see* HUMAN BIOLOGY & MEDICINE p.7) served as chief physician here from 1609 to 1643, during the same period that he was the guiding spirit of the College of Physicians. A clinical ward in the hospital and a small building are named after Harvey. (St. Bartholomew the Great, the City's oldest and perhaps most interesting church, is close by, but the hospital had its own small parish curch — St. Bartholomew the Less — still standing inside the gates on the left.)

GRESHAM COLLEGE (Barnard's Inn, Holborn)

Thomas Gresham (1519–1579), financier and founder of the Royal Exchange (*see* above), left instructions in his will for the erection of a college on property he himself had owned, and he also provided stipends for seven professors to give lectures — one each day of the week — on astronomy, geometry, physic, law, divinity, rhetoric and music. His legacy, Gresham College, became 100 years later the place where the Royal Society held its weekly meetings, and where some of the earliest members of the Society held professorial appointments: Christopher Wren, Robert Hooke (the Royal Society's salaried experimentalist), Samuel Pepys and others. The importance of this cannot be overstated; London was a crowded city and (in the absence of direct royal patronage) the lack of a congenial or even a tolerable headquarters might well have nipped the Society in the bud.

The original site of the college (Gresham's mansion while he was alive) was on Bishopsgate Street, just south of

present Liverpool Street station. A picture, frequently reproduced, shows a spacious quadrangle, with plenty of room
for Robert Hooke and others to have their homes and
offices and often their laboratories, but nothing remains
today — the area has become a huge modern office complex. Likewise a subsequent site on Gresham Street (named
for the gentleman in question), not far from the Guildhall,
has been put to other use. But, wonder of wonders, the
College itself survives, the lectures in the seven subjects are
still given (though not on a daily basis), and an eighth subject, commerce, has been added. It is now in Barnard's Inn,
a former Inn of Chancery and school for law students; the
building dates back to the 14th century and has been partially restored by the Mercer's Company — a delightful
place to see.

Lectures are free and normally open to the public without need
for tickets. Details are available from the College office. Phone
(0171)-831-0575. It may be noted that Gresham's tomb can
be seen at the altar of St. Helen's Church, off Bishopsgate on a
little street called Great St. Helens.

CHARTERHOUSE (Charterhouse Square)

Thomas Burnet (*ca*1635–1715), born in Yorkshire and educated in Cambridge, is famous for a book, *The Sacred
Theory of the Earth*, which has had an enormous influence
on geology, out of proportion to its scientific merit, today
regarded as minimal. Perhaps the title itself is its major
value, emphasizing, as it does, the basic need for a physical
history of the earth's surface as we see it now. The book's
ostensible goal was reconciliation of the biblical story with
such a history; Noah's flood was a central element, but (per-
versely, in the modern view) Burnet did not mention fossils,
which so many writers considered as prime evidence for the
deluge. The theory was criticised and attacked almost as
soon as it was published and has been a subject for debate
ever since — it is one of three central elements in a popular
book, *Time's Arrow, Time's Cycle*, by Stephen Jay Gould,
published in 1987!

The Charterhouse, originally a monastery, became a hospice and retirement home for pensioners in 1611; it still serves the latter function, being currently a residence for 40 pensioners, who (among other qualifications) must be bachelors or widowers and members of the Church of England. Burnet was made Master of the Charterhouse in 1685 and lived here for the rest of his life. He is buried in a vault in the Chapel.

The Charterhouse "campus" — almost completely walled off and hidden from public view — is entered from Charterhouse Square. Conducted tours for visitors are provided on Wednesdays at 2.15 p.m., April to July only. The public is welcome to attend services in the Chapel on Sundays (9.45 a.m. and 5.45 p.m), except in August. Phone (0171)-253-9503.

Plaque.   100 years after Burnet, there were still ardent advocates of the literal truth of Noah's flood. One of them was James Parkinson (1755–1824), physician, author of _Organic Remains of a Former World_. He is of course much better known for being the first to describe the neurological disability that we now call Parkinson's Disease. A plaque marks the house where Parkinson lived at 1 Hoxton Square, north of Liverpool Street station, and actually just outside the limits of the "City".

2. From the Inns of Court to Westminster

The Inns of Court were places of residence as well as lawyers' offices, and science in the old days was often done at home. We proceed here from Gray's Inn to the Strand and Trafalgar Square, a distance of a little over a mile. It's another quarter of an hour to Westminster Abbey — there are plenty of buses if you get tired of walking.

Gray's Inn (Holborn).   A statue of Francis Bacon stands on the lawn in South Square, with inscriptions that list his political and other offices. Bacon, widely admired as an avant-garde philosopher with positive views about experimental science (*see* PHILOSOPHERS p.6), was in fact a man of many parts — a courtier and lawyer, who became James I's Lord Chancellor in 1618. He was a student at Gray's Inn and retained his chambers here until his death, though his primary residence was in St. Albans. The base of the monument lists the impressive number of offices that Bacon held at one time, which includes offfices at the Inn itself.

Gray's Inn, a welcome haven of refuge from the traffic noise all around, is entered from Gray's Inn Road, close to its intersection with High Holborn.

Barnard's Inn (Holborn).   Barnard's Inn, the present location of Gresham College, has been included in the previous London group (*see* above) so as to keep it together with earlier sites of the College.

Red Lion Square (Holborn).   This square, a short distance west of Gray's Inn, is where John Harrison lived from 1739 until his death in 1776 — a plaque marks the site. The Board of Longitude's panel of experts came to this house in 1765 to dismantle and study Harrison's ultimate design for a marine chronometer; they reported favourably and Parliament (with some reluctance) awarded half the longitude prize to Harrison. It took ten more years of effort to get the other half. (*See* GREENWICH OBSERVATORY, Outer London, p.87.)

ROYAL COLLEGE OF SURGEONS AND HUNTERIAN MUSEUM (Lincoln's Inn Fields)

The Royal College of Surgeons occupies a fine building with an Ionic portico on the south side of Lincoln's Inn Fields. Besides lecture rooms and administrative offices, it has one of the most remarkable museums in London, the Hunterian Museum, devoted to the specimens collected and prepared by John Hunter (1728–1793), supplemented by later additions. Hunter, born in Scotland, is known as the "father of scientific surgery". He led a private school of anatomy where he had many subsequently illustrious students, including Edward Jenner, the pioneer of vaccination and immunology. Hunter insisted on a thorough understanding of anatomy, meticulous examination of healthy and diseased specimens, and even comparative study of human and animal parts — this depth of vision is reflected in the museum's collection, which is beautifully organised and displayed. Sad to say, the College was severely damaged by bombs in 1941 and half of Hunter's collection was destroyed. Replacements have been made: Hunter's original specimens are now marked with black identification numbers, later additions have red numbers.

The Royal College of Surgeons was not chartered until 1800, the need to find a home for the Hunterian collection being part of the motivation. Surgeons before that time were part of the guild of Barber Surgeons and the College still retains a chandelier from the old Barber Surgeons' Hall. The College entrance hall has a commemorative plaque for Joseph Lister and a grand collection of busts and portraits. Richard Owen (_see_ EVOLUTION & ANTHROPOLOGY p.39) was conservator of the museum from 1836 to 1856, before he moved to the British Museum and founded its natural history branch in Kensington. Arthur Keith was conservator from 1907 to 1933, a period that included the infamous Piltdown affair in which he was a major participant — possibly _the_ major participant.

The College and museum are open weekdays from 10 to 5. It is important to note that automatic access is granted only to the medical profession and to medical students, but the word "medical" is normally interpreted to included related sciences. Visitors without a professional connection should make advance arrangements: Phone (0171)-405-3474. An Odontological Museum with a unique collection of dental specimens is attached to the Hunterian Museum; rules for admission are the same.

The Temple (south side of Fleet Street).   The Temple (originally the seat of the Knights Templar) comprises two Inns of Court, the Middle and the Inner Temple, stretching from Fleet Street to the Thames. Today it is a warren of lawyers' offices, but 200 years ago Garden Court was the residence and laboratory of Smithson Tennant (1761–1815, _see_ SELBY, North Yorkshire p.247), an accomplished chemist of independent means, who gave lectures here to friends with similar interest and formed a partnership with William Wollaston (*see* EAST DEREHAM, Norfolk p.204) for the purification of platinum. In the course of this Tennant himself discovered two new elements in platina ore, iridium and osmium, presumably here at the Temple.

Garden Court is at the river end of the area, accessible from the Embankment or from Middle Temple Lane.

Crane Court (north side of Fleet Street). Crane Court was a transient home of the Royal Society after it left Gresham College in 1710. The house was used during Newton's presidency. On 2 May 1715, Newton, Halley and other notable figures observed a total eclipse of the sun from the roof — the time of the eclipse had been accurately predicted by Halley. The actual building was at the foot of the street where we now see the rear of a monstrous glass office building, but, apart from this intrusion, Crane Court today is lined by law offices and retains much of the narrow seclusion of days of yore. There is no plaque.

King's College (Strand). King's College was opened in 1831 as a conformist counterbalance to the undenominational University College on Gower Street. Here (in 1953) was done the x-ray diffraction work on DNA — the data used by Watson and Crick to propose their famous double helical structure (*see* MOLECULAR BIOLOGY p.49). A plaque just beyond the College entry identifies the building where the work was done and names the participants, including Maurice Wilkins, the group leader, who shared the Nobel Prize with Watson and Crick in 1962, and Rosalind Franklin, who is a major figure in James Watson's controversial book (*The Double Helix*) on the discovery, pictured as unreasonably reluctant to let the Cambridge pair see her data. One of the less familiar names on the plaque is that of A.R. Stokes, a mathematician. According to Wilkins, he had understood that the diffraction pattern must indicate some kind of helical structure earlier than Crick did so, but felt no urgency to rush into print.

National Portrait Gallery (Trafalgar Square). The gallery

has portraits of practically everybody, from royalty to entrepeneurs. More than 100 scientists and engineers are included, many of them in rooms 14 and 16. Portraits of many modern scientists (some still alive) are found in room 34.

WESTMINSTER ABBEY

There are no "two cultures" in this place of national remembrance — scientists and other celebrities are memorialized side-by-side. Isaac Newton's tomb is one of the grandest in the Abbey. Lord Kelvin is buried at his side and also has a huge window in his honour. Altogether about thirty scientists are represented, the exact number depending on how one chooses to define "scientist". Some, like Newton and Kelvin, were actually buried here; others have memorials in the Abbey but are buried elsewhere. One scientist, geologist William Buckland (memorial in the south aisle) was actually Dean of Westminster from 1845 to 1856. Few of the memorials attempt an account of the basis for recognition. James Prescott Joule is an exception, described as "establishing the Law of the Conservation of Energy and determining the Mechanical Equivalent of Heat"; Charles Lyell is another, cited for "deciphering the fragmentary records of the earth's history".

Some Abbey funerals are noteworthy. Irish Archbishop James Ussher (*see* DUBLIN, Ireland p.313), not a scientist by modern standards, perhaps, but judged to be one in his time, was buried here in 1656, in the Chapel of St. Paul. It was the time of the Puritan Commonwealth, but Oliver Cromwell ordered that the Anglican funeral service, then normally forbidden, should be used in deference to Ussher's status. Newton's funeral in 1727 was attended by Voltaire and has been described by him — the body lay in state in the Jerusalem chamber (public access through the deanery) and was followed to its grave by all the Royal Society. The 1912 funeral service for surgeon Lord Lister (memorial in the north choir aisle) has been described by Sir William Osler — the Abbey was packed to the door with nurses, students and doctors and there were reserved seats for representatives from all over Europe. We know of no scientist who was refused admission, as Lord Byron was when his funeral cortège reached the door in 1824 — poetic genius eventually won out over moral objections and a memorial was erected for him in 1969.

Visitors should remember that the Abbey is a church and will not admit sightseers when a service is in progress. Note also that there is an admission charge to pass beyond the nave, which one must do to see the famous poets' corner and the majority of the scientists' memorials. The *Official Guide*, available from the shop outside the Abbey, is indispensable for anyone seeking to locate a particular person.

Craven Street, where Benjamin Franklin lived during his many years in London. A plaque marks the house, just beyond the Waldor Hotel on the left.

Plaques. There is a plaque at 10 Maiden Lane (just north of the Strand) for the French philosopher Voltaire, who lived here in 1727 and 1728, when he was exiled from his own country. He and his mistress, the Marquise de Châtelet, were largely responsible for introducing Newton's work to French physicists. (Voltaire attended Newton's funeral at Westminster Abbey, as noted above.) Another plaque to a man we might regard as a foreigner is the plaque for Benjamin Franklin at 36 Craven Street (south of the Strand, by Charing Cross station), but Franklin was not a foreigner at all when he lived here, from 1757 to 1762 and from 1765 to 1775, for the American colonies were then part of Britain and Franklin himself did all he could to maintain the status quo. Franklin was an active member of the Royal Society and did his well-known wave-stilling experiments (pouring oil on water) during his second stay in London.

3. Piccadilly and Vicinity

ROYAL INSTITUTION (Albemarle Street)

The Royal Institution (founded 1799) was a marvellous establishment for over 100 years and still retains a special place. It was the brainchild of Benjamin Thompson (Count Rumford), American-born adventurer, scientist, and crusader for the use of science as a social tool. A specific mandate at the beginning was "Bettering the Condition and Increasing the Comforts of the Poor". It was to be supported by private subscription and unconventional means were created to encourage the rich to part with their money, popular lectures to which _ladies_ were admitted being particularly successful. Competent scientists were hired for salaried positions — most unusual at the time. Thomas Young, Humphry Davy, Michael Faraday, John Tyndall, Lord Rayleigh, James Dewar and William Bragg are among the illustrious names who became associated with the "R.I." in this way, the shining star being Michael Faraday, not only for his phenomenal scientific work (*see* ELECTRICITY & MAGNETISM p.18), but also for his role in the promotion of public understanding of science. He was a great lecturer; he instituted the regular Friday Evening Discourses and the ever popular annual Children's Christmas Lectures, now seen throughout the country on television. Faraday himself gave the children's lectures 19 times, at a guinea per head for adults and half that price per child.

Membership of the Royal Institution is open to anyone with an interest in science (no professional qualifications necessary) and members have access to the library and other parts of the building and can obtain tickets to the lectures. For the general public, however, there is only a small museum in the basement, a replica of Faraday's "magnetic laboratory", restored to its original size and condition. There is also an upstairs gallery, with a more comprehensive view of the Institution's research activities in the form of display cases devoted to Count Rumford, Thomas Young, and their successors. This gallery used to be part of the museum, but was sadly no longer open to the public on our most recent visit. (Visitors with sufficient desire to see it can probably get access by applying in advance. If so, given that the gallery surrounds the main auditorium, they may get a

Michael Faraday giving the Christmas lecture at the Royal Institution in 1855. Prince Albert is in the front row with the future King Edward VII beside him.

glimpse of the auditorium itself, with the original lecture demonstration table used by Faraday, and, in the anteroom, paintings of famous lecturers in action.)

The museum is open Mon–Fri, 1–4. Phone (0171)-409-2992.

BURLINGTON HOUSE

Burlington House, facing Piccadilly, was built in 1869 to 1873 specifically to house learned societies; present occupants include the Chemical Society, the Royal Astronomical Society, the Linnean Society, and the Geological Society. *Old Burlington House*, on the north side of the inner quadrangle, is occupied by the Royal Academy of Arts and usually thronged with visitors attending one of its exhibitions.

Linnean Society of London.   After the Swedish naturalist Carl Linnaeus — the man who devised our binomial system of classification — died in 1778, his heirs showed little interest in looking after his extensive specimen collection and library. They were bought in 1784 by the public-spirited Norfolk-born naturalist James Edward Smith. He became one of the founders of the Linnean Society (in 1788) and its president for the society's first forty years; when he died, the collection was purchased for the society from Smith's widow and is preserved by them to this day. The society's focus is on biological diversity and evolution; it publishes books and periodicals and holds regular meetings for its members. The most famous such meeting took place in 1858, when the society rooms were still in Old Burlington House — this was the meeting at which the papers of Charles Darwin and Alfred Wallace were read (*see* EVOLUTION & ANTHROPOLOGY p.37), announcing their essentially identical proposals for the evolutionary origin of species.

The society's library is open to the public on weekdays. A display cabinet at the foot of the stairs contains Robert Brown's microscope — the one he used for most of his observations and presumably the instrument through which he observed the phenomenon of Brownian motion (*see* MONTROSE, Scotland p.295). The nearby portrait of Linnaeus also belonged to Brown and was donated by him to the society.

Geological Society of London. This society was formed in 1807, meeting originally in Somerset House on the Strand. Its creation was at first controversial, perceived by many as an unwarranted intrusion on traditional turf of the Royal Society. It has an intriguing meeting room with opposing benches, as in the House of Commons, rather than the standard auditorium structure — quite appropriate in view of the many acrimonious professional arguments that have taken place here. There is a bust of Charles Lyell in the library and pictures of other famous geologists hang on the walls. One painting on the staircase is a depiction of some of the principal figures in the Piltdown case, in the process of examining the famous skull. This building is not open to the general public, but anyone with even a casual professional interest can usually be admitted to the Library without advance notice.

The Linnean Society library is open weekdays from 10–6. Entrance is on the left under the arch through which one enters from Piccadilly; the Geological Society is opposite on the right side. Phone: Linnean Society (0171)-434-4479; Geological Society (0171)-434-9944; Geological Society Library (0171)-734-5673.

JERMYN STREET (Geological Survey)

The British Geological Survey was created as a government agency in 1835 with the full support of the prominent geologists of the time, for the collection of information rather than the promulgation of theories was perceived at the time as the pressing need. Henry de la Beche (1796–1855), one of the most avid collectors, was the first director and insisted right from the start that a museum for the display of rock specimens must be a part of the project, creating the need for spacious quarters. The first location (Craig's Court off Whitehall) soon became inadequate and the operation was transferred to a prime site at 27/28 Jermyn Street, one block south of Piccadilly, where Simpson's clothing store is now located. The building also housed a School of Mines and rooms for its professors; it had a large lecture room where

popular lectures were given until the end of the century —
geology and mining captured the imagination of an era
when practical science and its industrial exploitation were
deemed a key to future progress and prosperity.

The headquarters of the Survey are now in Keyworth near
Nottingham, but a London Information Office is maintained in
the Natural History Museum in Kensington, where maps and
books can be purchased and where advice on matters of environmental concern can be obtained. Phone (Monday to Friday)
(0171)-589-4090.

87 Jermyn Street. A plaque commemorates Isaac Newton's residence. _See_ LEICESTER SQUARE, below.

The Royal Society Today (Carlton House Terrace). The
founding of the Royal Society in 1662 and the launching of
its _Philosophical Transactions_ in 1664 were momentous
events in the history of science. The institution still flourishes, the _Transactions_ are still published, but the organisation has grown as science itself has grown and has become
transformed in the process — directed more towards public
policy, less towards the individual scientist. Its official
duties today include giving scientific advice to the government and its principal role in the scholarly community has
been reduced to that of an honour society — it is a mark of
scientific prestige to be "FRS"; an embarrassment for some
to remain excluded as they approach the end of their career.
It is sadly no longer a place for discussion of fundamental
truths. Can you imagine any present member of the Royal
Society (or its American equivalent, the National Academy
of Sciences) setting up a laboratory experiment at a meeting
in order to demonstrate his latest finding to his colleagues?
Or a modern membership committee repeating experiments
at its sessions to decide whether some foreign researcher
merits admission?

Present quarters are in a stately home on Carlton House
Terrace, up the broad steps from The Mall and Buckingham
Palace. Walls of meeting rooms and stairwells are covered
by a wonderful collection of portraits, including Charles II
and some other founding fellows and even a few foreigners
who were not members at all. (The portraits were acquired
mainly by gift; they are not the product of a solemn selec-

tion process.) The building is not open to the public — to allow access one day a week or so might be good public relations for the scientific community.

Isaac Newton at Leicester Square. Isaac Newton moved to London from Cambridge when he became Warden of the Mint in 1696. A residence was provided for him in the Tower, adjacent to the Mint, but proved to be intolerably noisy and smoky (*see* p.56 above). Newton moved away as soon as he could, to the greener area of what was known at the time as Leicester Fields. Apart from one brief interval of a few months this area was his home for 29 years. Two houses are involved: the site of the first at 87 Jermyn Street (1696–1709) is marked by a plaque, the second (1710–1725) was on St. Martin's Street, a location now occupied by the Public Library of the City of Westminster. The second house was substantial: Newton shared it with his niece (Catherine Barton) until her marriage in 1717 and he had many servants. He moved away with some reluctance in 1725 (not long before his death), to Kensington in a quest for even more fresh air.

A "Newton Gate" at the southwest corner of the small park area in Leicester Square (quite close to the second house) celebrates his former presence here with a fine bust on a pedestal. There is also an explanatory plaque, but it's *not* about Newton: it informs us instead about the City of Westminster's continuing efforts to provide more park areas like this one.

32 SOHO SQUARE

Soho Square, close to Tottenham Court Road Underground station, is about 15 minutes' walk from Piccadilly Circus. Number 32, in the southwest corner, was the London home of Joseph Banks (1743–1820), the leader of the party of scientists who accompanied James Cook on his first voyage of discovery — the name Botany Bay for the expedition's first anchorage in Australia reflects the spirit of the voyage. The house was huge, an institution more than a residence. It provided rooms for Joseph Banks's associates, among them Daniel Carl Solander (1736–1788) and Robert Brown

(1773–1858), and for the specimens and drawings from the Cook voyage plus those collected subsequently from other travellers. Brown was responsible for the eventual transfer of the collection to the British Museum and later still it formed the nucleus of the present Museum of Natural History on Exhibition Road. Note that Banks was president of the Royal Society for an unprecedented 42 years (1778–1820). Brown was president of the Linnean Society, which held its meetings in this house from 1821 to 1857. This was truly England's nerve centre for natural history for close to eighty years!

Number 32 Soho Square is now occupied by a modern building, housing the 20th Century Fox Company. Neon lights advertise the fact, but there is also a tablet in the wall for remembrance of Joseph Banks and his associates. Soho Square used to have many grand mansions, only the House of St. Barnabas-in-Soho in the southeast corner retains its original character.

Broadwick Street (Soho).   Here we have the site of a celebrated water pump! London physician John Snow (1813–1858), with offices nearby at 54 Frith Street (where there is a plaque), demonstrated in this place that public water supplies can be the source of infectious disease — he traced 500 fatal cases of cholera in 1854 to water from this single source, and halted the epidemic by taking the handle off the pump. Note that this was well in advance of Pasteur and Koch; before micro-organisms had been identified as infectious agents. The "John Snow" pub now occupies the site, with a replica pump outside and an account of Snow's work on the wall of the upstairs bar. (John Snow was a versatile man. When the anaesthetic properties of ether and chloroform were first recognised, he devoted himself to the development of techniques for precise administration. He was Britain's first professional anaesthesiologist.)

4. British Museum and Gower Street

BRITISH MUSEUM (Great Russell Street)

This is one of the great museums of the world, but mainly dedicated to the arts and civilizations of past ages.

One science-related item is the Rosetta Stone, displayed in the Egyptian Sculpture Gallery on the ground floor. It contains the trilingual text — hieroglyphic, demotic and Greek — of a decree issued on the first anniversary of the coronation of Ptolemy V, King of Egypt. The text, finely chiselled on a slab of black basalt, is still clearly legible except where the stone itself has been damaged. Posters mounted beside the stone give an account of its history and the Anglo-French rivalry that was a part of it. Officers of Napoleon's French army made the discovery and recognised its importance, but the stone came to England as one of the spoils of war. Thomas Young, the great polymath of British science around 1800 (first professor at the Royal Institution), vied with the Frenchman Jean Champollion in the decipherment of the hieroglyphics. Young was the first to recognise that some hieroglyphics were alphabetical characters in spite of their pictorial appearance; he published all his evidence, but Champollion stuck for some time with the more conventional view that they were all pictorial symbols. Young, in turn, proved to be wrong in many of his specific assignments and Champollion, eventually won over to the alphabetical theory, is credited with the ultimate definitive transliteration of the text. (However, it took a new bilingual text, discovered later in another place, to convince Champollion and set him off in the right direction — he never acknowledged Young's priority for the basic underlying *idea*.)

The British Museum is open all year, Mon-Sat 10-5, Sun 2.30-6. Phone (0171)-636-1555.

UNIVERSITY COLLEGE (Gower Street)

University College was founded in 1826 as a non-denominational institution, by friends of religious liberty, who objected to the rigorous exclusion of dissenters from Oxford and Cambridge. It ultimately became merged with other London colleges into London University. It has a distinguished record in science.

Chemistry: Christopher Ingold building. This modern building, with entrance on 20 Gordon Street, has an exceptionally complete historical poster display on the ground floor. The wealth of talent that has occupied professorial seats in the department is truly impressive — the following are outstanding examples:

Alexander Williamson (1824–1904), professor here from 1855 to 1887, was an organic chemist, with a degree obtained with Liebig in Giessen. He revolutionised the way organic chemists think, a revolution that had its origins in attempts to synthesise higher alcohols (longer hydrocarbon chains than ethyl alcohol). For example, Williamson expected to make butyl alcohol by reacting the Na or K salt of ethyl alcohol with ethyl iodide: NaI or KI would be formed and a second ethyl group would add to the already existing one in the ethylate. To his surprise, his product was an ether instead of an alcohol, specifically (in this case) diethyl ether. This famous synthesis is a landmark, not so much for its own sake, but because it forced chemists to think about their reactions in dynamic terms, dominated by movement of *groups of atoms* (now universally called "groups") from one end of a molecule to another or between molecules. From here it was just a short step to representing the formulas of organic molecules in terms of such groups and to mentally equate the letters for the elements with actual atoms and the formulas for the way they are connected. Thus diethyl ether is $C_2H_5OC_2H_5$, whereas Williamson's originally expected product (same overall composition) would be written as C_4H_9-OH.

William Ramsay (1852–1916), Williamson's successor, professor from 1887 to 1913, is one of the most outstanding figures in the entire history of chemistry. He and Morris Travers and other collaborators discovered, isolated, and first prepared in pure form the entire family of inert gas elements — He, Ne, A, Kr, Xe, and (finally in 1903) radon, the gaseous emanation in the radioactive decay of radium.

Ramsay won the Nobel Prize in Chemistry for this work in 1904. Ramsay was born in Glasgow, his first university appointment was there, his interest in inert gases was aroused there by a lecture given in 1875 by Norman Lockyer on the solar spectrum, which of course referred to helium, then thought to be an element not found on earth. But all his actual work on the subject was done after he came to University College. (The chemistry department's reading room is named after Ramsay and his portrait hangs within.)

Christopher Ingold (1893–1970), after whom the building is named, was professor from 1937 to 1961. He is celebrated for his electronic theory of organic reactions, originally regarded by his peers as the "English heresy", but later to become the standard dogma in the field. A display case on the stairs gives more insight into Ingold's work, with details of his use of spectroscopic data as an experimental aid. Much of Ingold's work was done in collaboration with a Welshman, E.D. Hughes, who himself became head of the department after Ingold, from 1961 to 1963.

Other familiar names in the chemistry poster display include Thomas Graham (head from 1837 to 1854), who came here after a distinguished career at Glasgow (*see* GLASGOW p.290); F.G. Donnan, thermodynamicist (head from 1928 to 1937); Kathleen Lonsdale, crystallographer, professor from 1950 to 1965; and G.S. Hartley, the first person to understand soap and detergency, who was lecturer here from 1932 to 1945.

Physiology. A most memorable discovery made at University College is the discovery of hormones — chemical messengers — by William Bayliss and E.H. Starling in 1905. They proved conclusively that the secretion of digestive enzymes was not directly mediated by the nervous system (as had been thought), but involved an intermediate substance, a "chemical messenger" — hence the name "hormone", from the Greek word for setting things in motion. An Institute of Physiology was erected shortly thereafter and the building still stands, seemingly hardly changed at all, a diverting contrast to the modernity of the Chemistry Building. It is now called "Medical Sciences" and is entered from an inner court between Gower Street and Gordon

Square. There is a commemorative tablet for Bayliss and Starling.

University College Hospital. The first operation in Europe using ether as an anaesthetic was performed here in December 1846; it was the amputation of a leg, and the surgeon was Robert Liston. The building where the operation was done stood at 52 Gower Street (where there is now a student residence), a little down the street from the present modern hospital complex. A plaque put up by the Association of Anaesthetists of Great Britain and Ireland marks the spot. (An interesting sidelight is that now, 150 years later, we still don't know how ether and other inhalation anaesthetics work; we don't know the mechanism of action. It is a very active field of current research.)

London School of Hygiene & Tropical Medicine. This famous institution at the corner of Gower and Keppel streets (unrelated to the University) is the progeny of the research of Ronald Ross and Patrick Manson into the etiology of malaria (_see_ HUMAN BIOLOGY & MEDICINE p.11). It was founded in 1899 and has been in its present location since 1929; its library is the definitive source of references on the subject. The institution has a plaque inside for Patrick Manson and its main lecture theatre is named after him, but the only memorial for Ross is a bust on a back corridor — no intended slight, but simply a reflection of the fact that Manson founded the school.

(Likewise, the headquarters of the Royal Society for Tropical Medicine at 26 Portland Place, near Oxford Circus, is called "Manson House" because the building was given to the society by Manson. Ross actually lived just a few steps from Portland Place, at 18 Cavendish Square, and here there is a plaque in his honour, put up by the Greater London Council. It cites Ross as "Discoverer of the mosquito transmission of malaria", without reference to Manson.)

Other plaques. The present Biological Sciences building on Gower Street stands on a site where Charles Darwin had his home from 1839 to 1842. There is a blue plaque outside and a wooden one within. The economist John Maynard Keynes — if not the founder of scientific economics, he was a necessary precursor — was an adherent of the Bohemian Bloomsbury group of writers. A plaque marks the house where he lived at 52 Gordon Square, just a few steps from the University College chemistry building.

WELLCOME INSTITUTE (Euston Road)

The Wellcome Institute is the principal organisation in Britain for the support of the history of medicine. It has a fine library, open on application to anyone with an interest in the subject. It sponsors the exhibiton on the history of medicine at the Science Museum in Kensington, but also has a smaller permanent exhibition here on Euston Road.

The museum is open Mon–Fri 9.45–5; Sat 9.45–1. Phone: (0171)-611-8888.

5. Kensington Museums to Paddington

The South Kensington area was created as an exhibition area by the Great Exhibition of 1851, inspired in large part by Prince Albert, Queen Victoria's consort. Today it includes the Science Museum and the Natural History Museum, world-famous examples of institutions that may be described as *comprehensive museums* — giving an encyclopedic kind of overview, but lacking the *local* association that we stress in this book. Across Kensington Gardens, in Paddington (near the great railway station designed by Isambard Kingdom Brunel, also in the early 1850s) we have a contrasting kind of museum, more intimate, focused on a single scientific discovery and located in the actual building where the discovery was made.

SCIENCE MUSEUM (Exhibition Road)

The Science Museum has a remarkably comprehensive

scope. It includes the eye-catching products of technology that every museum of this kind must show — Stephenson's "Rocket" locomotive, a 1905 Rolls-Royce, a reconstruction of the Apollo 10 command module and lunar lander, for example. But the overall collection, augmented by dioramas and posters, manages also to give an overview of pure science: all fields are represented, virtually every advance or discovery mentioned in our book makes its brief appearance with a few words and a picture or two. It is too much to take in on a single visit and we note that multiple visits are encouraged by a very modest price for an annual pass as compared to the one-time entrance fee.

Specific items include (on the ground floor) a modern version of Foucault's pendulum, less impressive than the original because of a quite small amplitude of pendulous swing. The second floor has a good exhibit on nuclear physics, the atom bomb and useful nuclear energy, including, for example, a picture of the Hahn and Meitner apparatus in the German Museum in Munich and an explanation of why Hahn did not at first appreciate what he had wrought. The chemistry rooms on the same floor show atomic models from Dalton's wooden spheres to the most modern representations; also a history of chemical industry in Britain, from the salt works in Cheshire and the early dye industry to modern times. The third floor has more on physics — heat and thermodynamics, electricity and magnetism, optics, etc. It includes Joule's original apparatus for creating heat from friction, one of a number of items that are in the Science Museum by default because museums closer to "home" (in this case Manchester) lack the interest or resources to display them.

Biochemistry and cell biology are well represented (on the second floor) with a large display on "Living Molecules" — DNA, proteins, antibodies, muscle filaments, etc. An unusual item is a Svedberg ultracentrifuge, a huge instrument in which solutions containing large molecules are exposed to very high centrifugal fields in cells in which the resulting outward sedimentation can be followed by optical methods. (The characterisation of macromolecules that can be obtained by this means has loomed large in the biophysical research done by the authors of this book during their days in the laboratory — we admit to some nostalgia on seeing Svedberg's original design.)

The fourth and fifth floors of the museum, smaller in area than the others, house the Wellcome Museum of the History of Medicine. It includes forty tableaux giving glimpses of

William Herschel built and sold over 100 telescopes — this one can be seen in the Science Museum.

medical history from neolithic times to the present — showing a model of the Padua anatomy theatre, for example, and such scenes as Joseph Lister's having disinfectant sprayed around his operating table. An upper gallery has a somewhat more systematic historical orientation, with successive periods ("Tribal Society", "Oriental", "Mesopotamia and Egypt", etc.) each given its own specific section.

There was a gimmicky feature we didn't like: the presence of actors dressed as scientific personalities (e.g., Faraday) and pretending to speak with their voices.

The museum is open Mon–Sat 10–6, Sun 11–6. Phone (0171)-938-8000 or 938-8123 (recorded information). The museum library, devoted to the history and public understanding of science, is around the corner on Imperial College Road and is open to the public. Phone (0171)-938-8234.

NATURAL HISTORY MUSEUM (Cromwell Road)

The Natural History Museum and the Geological Museum used to be separate entities but are now combined; the main building is an elaborate architectural structure, worth viewing for its own sake. The exhibits remain divided more or less as they were when the museums were separate. The "Life Galleries" include dinosaurs, our place in evolution, human biology, marine invertebrates, "creepy-crawlies", etc. The "Earth Galleries" contain a good exhibit called _Story of the Earth_, but without names of people or any reference to the titanic struggles that often took place to establish individual chapters of the story — which is a pity because many of the controversies make fascinating reading. The exhibit does, however, accurately present the current views on the origin of sun, earth and moon, describes the inner core of the earth, the surrounding mantle and the outer crust; it has up-to-date accounts of modern topics, such as reversal of magnetic field and plate tectonics. Overall, the museum leans toward the trendy type of presentations (the dinosaurs are animated), ostensibly to lure an otherwise uninterested public.

A fascinating organisation in the museum's back rooms is the International Commission on Zoological Nomenclature, the body responsible for keeping track of the scientific names of all species of the animal kingdom (fossils as well as alive) and for assigning names to new species — 15 000 new species names are added to the zoological literature each year! The commission's quarters are not open to the public, but visitors with any legitimate interest are warmly received. There is also a "Classification" section in the main galleries, devoted to the subject.

An interesting story relates to the statue commemorating Richard Owen, the founder of the museum, which stands impressively on the main staircase. Owen had for many years been supervisor of the natural history collection of the British Museum and presided proudly over its move to independent quarters in Kensington. But he was also a bitter opponent of Charles Darwin's theory of the origin of species (though not of evolution per se) and thereby incurred the enmity of Thomas Huxley, the influential pro-Darwinian crusader. Huxley contrived to have a statue of Darwin placed on the staircase when the museum was first opened, to the chagrin of the staff as well as Owen himself. The insult was quickly repaired: Darwin's statue was removed and eventually replaced by the one we see now.

The museum is open Mon–Sat 10–6, Sun 11–6. Phone (0171)-938-9123. An Information Office of the British Geological Survey is maintained in the museum (*see* PICCADILLY, p.71 above). Visitors need no appointment and can obtain passes at the museum entrance without paying the normal museum entrance fee. Phone (Mon–Fri) (0171)-589-4090.

CHELSEA — CARLYLE'S HOUSE (Cheyne Row)

Thomas Carlyle, opinionated writer, famous (or infamous) for his convoluted style, moved from Edinburgh to London in 1824 and at once complained: "What a sad want I am in of libraries, of books to gather facts from. Why is there not a Majesty's library in every county town?" In 1837 there were indeed only 240 000 volumes in the British Museum library and they were in total disarray, mixed up with collections of fossils and other artefacts. Carlyle set about establishing a private library for use of London's intellectual community. He lived and entertained at his house in Chelsea, and one fascinating point about his guests is the mixture of literati and scientists. Dinner guests included everyone who was anyone. Charles Darwin and Charles Dickens might well sit at the same table. Lyell was a frequent guest, as was Charles Babbage, retrospective "father" of computers. There was much verbal sparring and acrimonious argument and Darwin, in particular, had little use for Carlyle: "I never met a man with a mind so ill adapted for scientific research," he wrote in his autobiography. Yet this mix of dinner guests was surely an advantage for original thought — what could be deadlier than a company of men who all think alike?

Carlyle's house has been well preserved and successfully evokes the genteel comforts of London life at a time when so many scientists lived and worked in the city. (Note: We cannot be entirely sure that Charles Darwin actually ever came to Carlyle's house. Charles's older brother Erasmus was the family enthusiast for dinner parties and would certainly have attended frequently; Charles and Emma in turn attended dinners at Erasmus's house and that is where the encounters with Carlyle may have taken place.)

The house is open to the public Wed–Sun, April to October. Phone (0171)-352-7087. The house has electricity only for minimal custodial functions. Note also that the "comforts" of the day did not yet include flush toilets.

Chelsea Physic Garden (Royal Hospital Road). South of the museums, between Royal Hospital Road and the Chelsea Embankment, lies the Chelsea Physic Garden, founded in 1673 by the Society of Apothecaries to culture and study the therapeutic properties of plants. They have been doing so ever since and have opened their garden to the public on Sunday and Wednesday afternoons, April to October.

An interesting historical note: there used to be bitter rivalry between physicians and apothecaries, because the latter had the right to _prescribe_ medicines as well to compound and dispense them. William Watson (Fellow of the Royal Society and active member of London's scientific establishment) had begun his career as a botanist and apothecary, but around 1755 wanted to start a medical practice. The physicians had an absolute rule against admitting apothecaries to their ranks, so that Watson first had to resign from the Society of Apothecaries, which was not as easy as one might think — Watson was charged a £50 resignation fee, a huge sum in those days.

The entrance to the garden is on Swan Walk, on the east side. Open hours are April to October, Sun and Wed, 2–6. Phone (0171)-352-5646.

42 Rutland Gate. This house, mid-way between the Science Museum and Hyde Park, has a plaque honouring Francis Galton (1822–1910), who lived here for 50 years. Galton (born in Birmingham) was a grandson of Erasmus Darwin and cousin of Charles Darwin, and an intellectually precocious and versatile contributor to 19th century science. Meteorology was his first interest: he realised that air circulates in the opposite direction in high and low pressure centres, and coined the term "anticyclone" for the highs on the weather map. Publication of his cousin's *Origin of Species* made him an enthusiast for applying the concepts of evolution to human life — he did scholarly work to indicate that "talent" (intellectual, athletic, etc.) is inherited and advocated "eugenics" (selective breeding) to improve the human race, for which he would be pilloried today, of course. He studied the question of whether criminal tendencies could be recognised by body dimensions, but found the answer negative: a useful byproduct was the demonstration that fingerprints are unique (non-discriminatory) attributes of each individual and it was Galton who first proposed their use to identify criminals in police records. In the course of his work on human mensuration he developed statistical methods that have become textbook material: he originated the concepts of regression and correlation, for example. He seems to merit more admiration than he gets, but presumably his advocacy of eugenics makes this difficult in today's unforgiving world.

PADDINGTON —
ALEXANDER FLEMING MUSEUM

The Scottish bacteriologist Alexander Fleming (1881–1955) is one of the heroes of 20th century medicine for his 1928 discovery of penicillin, which spawned the host of antibiotics that have made death from bacterial infection almost a thing of the past. The discovery was made at St. Mary's Hospital, just a few steps from Paddington station, in a tiny laboratory which has been preserved as a monument and museum. The laboratory gives a realistic impression of the austere conditions under which so many great discoveries used to be made. The discovery itself is a vivid illustration of one of the fairly common routes to knowledge — an accidental observation, coupled with a prepared mind that could be alerted to its possible significance.

What happened was so simple that even the lay person can appreciate it without difficulty. Fleming had been away on holiday and was clearing away some petri dishes with bacterial cultures (_Staphylococcus_) that he had been studying. He found that one of the dishes had become contaminated by a mould, a not uncommon experience in any laboratory. This one happened to be green in colour, but what struck Fleming in particular (and lesser minds might never have noticed at all) was that the bacterial culture in a wide circle about the green spot had been dissolved. "That's funny," he said as he went on to investigate. He identified the mould as belonging to the _Penicillium_ family; a team of organic chemists at Oxford (headed by Howard Florey and Ernest Chain) went on to purify and study the bactericidal substance itself; the first clinical test was performed in 1941 at the Radcliffe Infirmary in Oxford. A crash programme of large scale production of penicillin was initiated in the U.S. for use on World War II battlefields. Fleming, Florey and Chain shared a Nobel Prize in 1945.

Alexander Fleming came from a humble Scottish hill farm family, but he was an experienced bacteriologist by 1928, who had already discovered a bactericidal agent, albeit one of very limited applicability — the protein enzyme lysozyme, a component of human nasal mucus and tears. He is not St. Mary's only Nobel laureate, for Rodney

Porter, immunologist then at St. Mary's, won a share of the prize in 1972 for his work on the structure of antibody proteins.

St. Mary's is on Praed Street, on the east side of Paddington station; the entrance to the museum is on Norfolk Place, just off Praed Street. The museum is open Mon–Thu, 10–1, and other times on any weekday by appointment. Only a small number of visitors can be accommodated at a time and larger parties are asked to make appointments regardless of when they wish to come. Phone: (0171)-725-6528.

More plaques. James Clerk Maxwell lived at 16 Palace Gardens Terrace, W8, just west of Kensington Gardens, during the period (1860–1866), when he was professor at King's College. About one hundred yards further west is Kensington Church Street, where Isaac Newton lived the last two years of his life, until his death in 1827. Did Maxwell know that? The Newton residence was demolished in the 1890s and is now occupied by Bullingham Mansions (no plaque).

Another plaque is at 7 Kensington Park Gardens (Notting Hill, W11), where William Crookes (1832–1919) lived from 1880 until his death. His best-known work is his invention (around 1875) of the "Crookes Tube", a highly evacuated tube with sealed-in electrodes, which produce the so-called "cathode rays" at high voltage, later found to be streams of what we now call electrons — when they strike the anode they generate x-rays, as Wilhelm Röntgen discovered in Germany in 1895. But Crookes himself played no part in these discoveries. He was a mercurial character who never stuck to one thing for very long: by 1880 he was thick in controversial *psychic research* and in 1897 he was president of the Society for Psychic Research.

6. Outer London

DOWN HOUSE (Downe, Kent)

Down House, Charles Darwin's home, is formally within

the boundaries of Greater London, but historically and practically belongs in the county of Kent. We have listed it under Southeast England p.104.

GREENWICH OBSERVATORY (London SE10)

The construction of the Royal Observatory was ordered in 1674 by King Charles II to solve the problem of determining longitude for ships at sea. He said "let it be done in royal fashion"; he wanted none of his ship-owners and sailors deprived of any help the heavens could supply. As it turned out, the problem of making sufficiently accurate astronomical observations from the unsteady deck of a ship proved too difficult, and eventually it was an alternative method based on precise time-keeping that won out. The scientific basis for the latter is the familiar difference in time between two places, which is in direct proportion to the difference in longitude. Local time on board ship is easy to measure by using sunrise or sunset to calibrate one's watch, but where do we get an _accurate_ reference time with which to compare it? Parliament offered a prize of £20000 for the solution to the problem in 1714, but the prize was not awarded until 1773, when a Yorkshireman (John Harrison) proved able to construct a chronometer that would not deviate for weeks on end from the chosen reference time, which we now call "Greenwich mean time".

The smoke and street lights of the city have forced the working "Greenwich Observatory" to move away into the countryside, but happily the historic buildings have been preserved and now constitute a part of the National Maritime Museum, which is arguably the single most spectacular scientific tourist destination in all of England, the one museum that everyone should visit.

One of the old buildings is Flamsteed House, which was designed by Christopher Wren for the first Astronomer Royal and has been authentically restored. (Its ground floor and basement served as residence for the Astronomers Royal until 1948.) Of equal interest is the Meridian House added later — the Greenwich meridian is _formally and arbitrarily_ defined by the center of the astronomical transit

Flamsteed House, designed by Sir Christopher Wren. The ball on the turret (red in colour) was added in 1833. It was the world's first visual time signal, and is still raised and dropped at precisely 1 p.m. each day. The line across the picture from left to right (a little hard to see) is the Greenwich meridian.

instrument in this building and its position is marked on the courtyard ground. Historically there have actually been two meridians, one corresponding to the transit instrument of James Bradley (used from 1750 to 1850) and the other to the instrument installed beside it in 1851 by George Airy, the seventh Astronomer Royal. The lines are 6 yards apart, which is equivalent to 0.02 seconds of transit time, too small to be significant in Airy's time; it is the newer Airy line which is the modern international reference.

The two buildings together form the "Old Royal Observatory", on top of the hill in the park, and they represent the science-oriented part of the museum, dealing with the work of Flamsteed, Halley and their successors, and displaying the tools of the trade. As expected, the major emphasis is on the reading of celestial position and accurate clocks. Especially impressive are the huge protractor-like calibrated devices for measuring elevation above the horizon (quadrants, sectants, octants). A whole room is devoted to Harrison's chronometers and it includes the prize-winning chronometer H4, still in working order. Historical plaques are everywhere. Their reports of the quarrels between Flamsteed and Halley are particularly fascinating, good soap opera stuff — when Halley was appointed (age 64) to succeed Flamsteed in 1720, Mrs Flamsteed removed all astronomical instruments along with the furniture — Halley inherited a building with bare walls, which he had to re-equip at his own expense.

What are formally the main galleries of the Maritime Museum are actually downhill from the Old Observatory, closer to the River Thames. Here we find ship models, navigational instruments and sea charts; items relating to the explorations of Captain Cook; and much patriotic material about Britain as a seapower. The buildings are also of architectural interest, designed by Inigo Jones and completed in 1635, before the Observatory was even contemplated.

Greenwich is 7 miles east of the city centre. One can travel there by train from Charing Cross station, but a better way is via the Docklands Light Railway (part of the Underground system), using a pedestrian tunnel under the Thames from Island Gardens station. The pleasantest way (at least in summer) is to take a boat down the Thames from Westminster Pier. The museum and Old

Observatory are open Mon–Sat 10-6, Sun 12–6. Hours are slightly shorter in winter. Phone (0181)-858-4422.

HAMPSTEAD (London NW3)

Sigmund Freud lived here for the last year of his life and Hampstead is keen to advertise the fact. Whether to regard Freud as a scientist or a quack is debatable, but there is no question about the influence of Freud's work (and fame) on orthodox psychological science, away from strictly measurable phenomena towards what we might call the "hidden

Sigmund Freud's statue in Hampstead.

agenda" of the human mind. Freud lived in Vienna for most of his life, but, being a Jew, was forced to leave in 1938. He found refuge in England and lived for the last year of his life in a house at 20 Maresfield Gardens, which is now a museum containing his famous couch and other mementoes. For true disciples, of course, it can never compare with the house at Berggasse 19 in Vienna, which is the ultimate shrine, where Freud developed his ideas and actually treated patients.

Perhaps more intriguing than the museum — given the unpredictable psychology of municipal planners — is the existence of a full-size statue of Freud, on the grass beside the public library at Swiss Cottage.

Maresfield Gardens is a few minutes' walk from Finchley Road Underground station (clearly signposted), which in turn is about half a mile from the busy traffic centre at Swiss Cottage. The museum is open Wed–Sun, 12–5. Phone (0171)-435-2002.

HAMPTON (Borough of Richmond)

Exact mensuration is an essential part of science, and here we have an extraordinarily accurate measurement from 1784 — a spot marked on an old cannon embedded in the earth, which is precisely 5.1902 miles from a similar spot at what used to be the opposite edge of Hounslow Common, but is now part of Heathrow airport.

What was the purpose? Primarily it had to do with navigation. British sailors charted their positions with reference to the Greenwich meridian, but the French had their own meridian line, passing through their own royal observatory in Paris. To avoid the need for duplicate determination of positions of lighthouses and other fixed points, it was obviously essential to establish the _relative coordinates_ of the two observatories. The practical way to do this is by the method of triangulation, dependent on the fact that knowledge of the length of the base and two angles determines all dimensions of a triangle. A series of triangles from Greenwich to Paris thus solves the problem, and the base of the first triangle determines the accuracy of all subsequent measurements, for angular measurements to a distant church steeple or other marker are easier. The task on the

English side of the Channel was entrusted in 1784 to William Roy (1726–1790), mapmaker extraordinary, and he chose Hounslow Heath for the location because it was a huge uncluttered space at the time. Roy devoted months to "reinventing" (his own word) measuring devices (steel chains, brass pulleys, glass rods, theodolite) that would give the requisite accuracy and his eventual result was confirmed some years later to within less than 3 inches! The corresponding difference in longitude between Greenwich and Paris was 2°19'42" and, with this number, the Greenwich and Paris definitions of longitudinal "zero" could be used interchangeably. The Greenwich meridian was adopted as the universal standard at an international conference held in 1884, the centenary of Roy's measurement.

It should be added that Roy was an ardent advocate for an accurate survey of all of Britain. He regarded the Hounslow baseline as equally important for that project as for the Paris longitude, which it proved to be, and he is considered to be the "pioneer" of the present Ordnance Survey.

The termini of the Hounslow base were marked in 1791 with old cannons embedded in the earth, muzzle upward, the "spot" itself in the centre of a cap screwed over the muzzle. Bronze plates provide historical details. One of the termini is here in Hampton, in a small housing estate named after William Roy, on a street called Roy Grove. The other is within Heathrow airport close to the northern perimeter road.

Hampton is 1.5 miles west of Teddington on the A312. Roy Grove is a cul-de-sac off Hanworth Road. The railway station at Hampton is less than a mile away.

HAMPTON COURT
(Borough of Richmond)

Michael Faraday knew no home other than his modest flat at the Royal Institution. When the need came for him to retire, Queen Victoria provided a house for him at Hampton Court Palace. It was not an unprecedented act, for Queen Anne had done the same for architect/scientist Christopher Wren 150 years earlier. Both houses survive, on Hampton Court Road, at the corner of the Green, right opposite the west entrance ("Trophy Gates") to the Palace grounds, and they carry appropriate inscriptions. The houses are substantial, not to be compared with the "grace and favour" apartments still in use today for pensioners of the crown.

Hampton Court Palace is 2 miles south of Hampton, easily reached by rail from Waterloo station — it is on a different branch line from that serving Hampton or Teddington.

HIGHGATE CEMETERY (London N6)

A magnificent symbol of Victorian secularity, Highgate Cemetery is best known for the graves of Karl Marx and many literary figures. Marx's tomb, with the inscription "Workers of the World Unite", has been a popular site of pilgrimage for the faithful — presumably their numbers are not what they once were. The cemetery is of interest to us because Michael Faraday was buried here when he died in 1867. His grave is marked by a simple tombstone, bearing only his name and the dates of birth and death. This simplicity was specified in Faraday's will and respected the tenets of the Sandemanian religious sect to which he belonged, a sect opposed to all forms of spectacle and ostentation — in fact, opposed to any religious practice not specifically recommended in the Bible.

After Faraday was interred here, a group of public-spirited citizens (chaired by the Prince of Wales) sought a more prominent memorial for Westminster Abbey and commissioned a statue for this purpose, but Faraday's family was adamant that no likeness of him should exist in a church that he would never himself have entered — the statue can be seen today in the entrance vestibule of the Royal Institution (_see_ PICCADILLY, p.67). A small plaque was eventually placed in Westminster Abbey in 1931.

Faraday's grave is in the western section of the cemetery. In contrast to the eastern section, access to this section is limited. There were daily conducted tours at noon when we last visited. Phone (0181)-340-1834.

KEW GARDENS
(Borough of Richmond)

Kew Gardens is one of the grandest botanical gardens in the world, mostly a showplace, of course, but also identified with the history of botanical science. The first gardener (when the place was still private royal property) was William Aiton (1731–1793) and he set the tone by his publication of a famous catalogue "Hortus Kewensis", at a time when works of that kind (other than for medically useful herbs) were still rare. The much celebrated Joseph Banks made Kew an international focal point for collectors from all over the world. Later (in 1840) the gardens were turned over to the nation, with William Hooker (1785–1865) as the first official director. He increased the size of the gardens from a mere 11 acres to near its present size of 300 acres; he used his own house ("West Park") to create a library and a herbarium, he created resources for practical applications — for example, *Cinchona* trees were brought in from Peru to provide a cheap source of quinine for the treatment of malaria. Research continues today at the herbarium (mostly taxonomic) and at the Jodrell Laboratory (chemistry and structure), first opened in 1876, when Hooker's son Joseph had become director. (The latter, besides his work at Kew, is also well known for his long friendship with Charles Darwin — it was Hooker and Charles Lyell who persuaded Darwin to agree to the joint presentation of Darwin's and Wallace's theories of evolution and natural selection to the Linnean Society in 1858.)

The outdoor areas of the Gardens present today's visitor with an ever-changing display of seasonal flowers and trees. Buildings with controlled environments, not subject to seasonal variations, include the Princess of Wales conservatory, with areas of 10 different tropical climates; the Palm House, a masterpiece of Victorian engineering, containing many rainforest species plus a marine display in the basement; a Temperate House for items like citrus trees; an Alpine House; and several other smaller buildings. A prominent building at the northern tip of the Gardens is named after Sir Joseph Banks and used for special exhibits: when we visited the exhibit was called "Thread of Life", sponsored by Courtaulds company and devoted to the many useful fibres made from plant products.

Kew Palace (north side of Gardens).   The present palace was once an annexe to a larger royal residence, now long demolished. A sundial in front of the building marks the former location of Kew House, the private residence of Samuel Molyneux (1689–1728), which stood here long before the Gardens were even planned. Molyneux, born in Chester and educated at Trinity College, Dublin, had an amateur's zest for astronomy and equipped the house with precision apparatus for making stellar observations — many of them carried out in collaboration with James Bradley (*see* WANSTEAD, p.98), later to become Astronomer Royal. In 1725 they discovered astronomical aberration (not to be confused with *optical* aberration of lenses), a phenomenon due to the speed of motion of the earth in its orbit around the sun, which affects the apparent direction from which the light of a star is coming. A little later Bradley made another important discovery in relation to astronomical observation, namely the nutation of the earth's axis, a periodic wobble superimposed on the very slow precession of the earth's axis of rotation. In retrospect these were sophisticated concepts — arising, as they did, from motions of the observer's platform — which provided essential corrections for all future astronomical data.

Kew Gardens is about the same size as Hyde Park and can be entered by several gates: Victoria Gate, near the centre of the park, is the closest to the Underground (Kew Gardens station); there is a car park at Brentford Gate (northwest corner). The Gardens are open daily all year, 9.30–4 in winter, 9.30–6.30 in summer. Annual season tickets are available at reasonable price for multiple visits. Phone (0181)-940-1171. The research buildings (herbarium and Jodrell laboratory) are not open to the public.

MILL HILL (London NW7)

Ridgeway House, now part of Mill Hill School, was the home of Peter Collinson (1694–1768), a businessman specialising in trade with the American colonies. But trade was by no means his only interest, for he accompanied it with an intense correspondence abroad, exchange of information and materials, etc. — for example, he had extensive gardens here at Mill Hill where he cultivated what to English eyes were exotic plants, sent to him by American merchants. Collinson was an active fellow of the Royal Society and a spokesman for its scientific activities outside the British Isles. Most famously, he was a patron for Benjamin Franklin — sent him (to Philadelphia) the instruments with which Franklin did his electrical experiments, sponsored the publication of his results in the *Transactions*, and (after Franklin came to England in 1757) welcomed him into the inner circle of Royal Society activities.

Collinson's own publications were sparse. One of them concerned the migration of swallows, contending that swallows migrate between summer and winter ranges, and do not — as everyone seemed to believe — survive the winter by hibernating under water near their nesting sites.

Mill Hill School is on the Ridgeway (nearest Underground station Mill Hill East) and Collinson's residence there is marked by a plaque.

TEDDINGTON (Borough of Richmond)

Stephen Hales (1677–1761) was the most influential of England's many research-minded vicars. He was "perpetual curate" here at St. Mary's Church for more than 50 years and took his parish duties seriously, but at the same time managed to become the acknowledged founder of plant physiology — he measured rates of water and air uptake and efflux, and demonstrated the importance of a component of air that was distinct from the rest, which was of course carbon dioxide (CO_2), still unknown at the time. The apparatus he devised to handle and measure gases in such experiments was in fact then used by Joseph Priestley and others for subsequent chemical research into the composition of air.

Hales's grave is in the churchyard, but there is also a monument within the church on the floor of the porch. It should be noted, however, that an actress, Peg Woffington, one of Hales's parishioners, is more heartily remembered in the parish than he is — her former home is a popular tea-room close to the church.

St. Mary's Church is at the intersection of Twickenham Road and Ferry Road, dwarfed by its partner church (St. Alban) across the street. It is normally kept locked except for services. For access phone the vicarage: (0181)-977-2767. A pleasant one-mile walk can be taken in good weather between Hampton Court Palace and Teddington, traversing Bushy Park, a green area filled with grazing deer. The National Physical Laboratory (closed to the public), first installed at historic Bushy House in 1902, is at the edge of the park.

UPMINSTER (Borough of Havering)

William Derham (1657–1735) was another committed practicing clergyman, rector of Upminster for nearly 50 years from 1689 until his death. He was a Fellow of the Royal Society, delivered the Boyle lectures at the Royal Society in 1711–12, and wrote books about John Ray and Robert Hooke and their work. His best known exploit is the measurement of the velocity of sound in 1705, which led to what is now regarded as a black mark on the record of Isaac Newton. Newton had made a crude measurement in the Great Court at Trinity College, leading to a value of 968 feet/sec, whereas Derham made a much more accurate measurement of 1142 feet/sec. The Derham measurement involved prearranged firing of cannons at half-hour intervals by a gunnery unit stationed at Blackheath (12.5 miles away). Derham could see the flash from the tower of the Upminster church and then could hear the sound a little later. His published report is an exemplary model of a careful experiment, including all manner of controls on the validity of the result, e.g., barrels of the guns were pointed towards Upminster in some firings but in the opposite direction in others without effect on the result.

No blame, of course, attaches to Newton's inaccurate experiment, based, as it was, on a very short travel path in comparison with Derham's 12.5 miles. What gives Newton a black mark is that he theoretically _predicted_ a value near 968 feet/sec in the first edition of _Principia_, and then changed the prediction in a later edition in order to arrive precisely at Derham's result. It wasn't really dishonest, because the parameters entering into the calculation were what we now call "guesstimates", but it was distinctly self-serving.

St. Laurence church in Upminster is still the parish church. The building is unremarkable, except for the crucial tower, which dates from the 13th century. It has a wooden pointed top, high enough to stand out over all the surrounding buildings. We didn't check to see whether Blackheath is still visible from it. There is a tablet on the wall inside the church at the base of the tower, listing all the rectors from 1336 onward. A modest framed notice beneath it draws attention to the life and deeds of William Derham, without specifically mentioning the measurement of the velocity of sound. One of the utilitarian outbuildings of the church is named Derham Hall and carries a similar inscription.

Upminster is 19 miles east of central London and is served both by Underground and British Rail. The station is close to the church.

WANSTEAD (London E11)

This is one of the rare cases where one can identify a unique source of inspiration for a famous scientist, a benefactor without whom no thought of a scientific career would have been conceivable. The famous man in this case is James Bradley, professor of astronomy at Oxford and the third Astronomer Royal; the benefactor is his uncle, James Pound (1669–1724), rector of St. Mary's Church at Wanstead. Pound was himself a reputable amateur astronomer; his measurements on Saturn's ring and satellites and diverse data on Jupiter were used by Newton and Laplace, for example. But more important in the long run was his fondness for his nephew, James Bradley, whom he trained and with whom he made many joint observations. Later, after Pound's death, Bradley set up his own telescope and sector here; the discovery of stellar aberration, based on observations begun at Kew with Molyneux, was actually conclusively established here. Bradley went on to succeed Halley as astronomer royal — he held the post for 20 years, with great distinction and brilliance (*see* GREENWICH OBSERVATORY, p.87).

St. Mary's Church was rebuilt in 1790 and Pound's grave, originally within the building, now lies in the graveyard outside, on a path leading south from the vestry door. A commemorative stone was placed beside the actual gravestone in 1910 by fellows of the Royal Astronomical Society and others.

Wanstead Park.   Wanstead used to have several grand mansions and Pound's second wife (late in life) was the heiress of one of the lesser estates. Another house (on the grounds of Wanstead Park) is where the Pound/Bradley telescope was actually set up. It was mounted on a maypole that had stood in London on the Strand and was procured for Pound through the influence of Isaac Newton — the objective lens was on loan from the Royal Society. The grand houses have all disappeared, but the park remains, since 1881 a public park with three attractive lakes.

The Parish Church of St. Mary is on St. Mary Avenue, close to Wanstead Underground station. Wanstead Park is a few minutes' walk beyond the church, in a bend of the River Boding. The church is closed when not in use. Phone (rectory) (0181)-989-9101.

2
Southeast England

ALBURY (near Guildford, Surrey)

The Old Parish Church in Albury is a fascinating place. It is
of venerable age, but was made redundant in 1842, when the
village was moved about a mile away to provide more
breathing space around the manor house, and the altar,
church bells, etc., accompanied the congregation to a new
location. Work of restoration, however, began in 1974; the
church was reconsecrated and services are now held there
twice a year. There are no pews, so worshippers must stand.
We visited just a week before one of the services and the
church was being filled with flowers in preparation for what
would clearly be a festive occasion.

Going back 350 years, the church had as its rector (for half
a century) one of Britain's outstanding early lights, William
Oughtred (*ca*1573–1660). He was a mathematician, tutor of
Christopher Wren, advocate of standard symbols for arith-
metic and algebra — the × for multiplication is one we still
retain. He is remembered above all as the inventor of the
slide rule, the indispensible *vade mecum* of every engineer
and chemist before the advent of the electronic pocket calcu-
lator around 1970. (Does the younger generation know what
a slide rule was? Napier's invention of logarithms made it
possible — logarithmic scales etched on wood or metal
whereby multiplicative products of numbers could be read
off in an instant.) The church has a memorial to Oughtred

within the building and a booklet for sale to decribe his work. He was buried in the chancel, but no inscription remains to indicate the exact location.

The church is also associated with Robert Malthus (1766–1834), who lived here with his parents (while holding the curacy of an outlying parish) in the period when he wrote his famous essay on population — setting out the principle that unchecked human population growth would eventually outstrip our food supply. There is an enclosure for family graves in the churchyard, but Robert himself is buried in the Abbey in Bath.

The church is in Albury Park, a mile west from the modern village, close to the intersection of the A248 with the A25. It is normally open in daylight hours. Phone (Albury Park administrator) (01483)-202964. Note that there is another (much larger) redundant church nearby — this one originally built for the Catholic Apostolic Church.

BURSTOW (near Horley, Surrey)

St. Bartholomy's Parish Church, in a tranquil wooded enclave on the fringe of this village (only 2 miles from Gatwick airport!), is the place where John Flamsteed (1646–1719), the first Astronomer Royal, was officially the rector for 35 years and where his body was taken for burial after he died at the Greenwich Observatory. His tomb is in the chancel, under a suitable memorial, and there is a nice poster on the astronomer's life and career next to the door that leads into the bell tower. They are the only formal memorial to Flamsteed — there is none in Derbyshire, where he was born. (It is worth noting that Burstow is about 25 miles from Greenwich, which made it possible, but difficult, to combine the duties of rector and astronomer. But curates actually did most of the church work in those rosy days and Flamsteed probably officiated himself only a few times a year.)

A familiar anecdote about Flamsteed's wife tells that she had no liking for Edmund Halley, her husband's successor at Greenwich, and that she removed every scrap of astronomical equipment from the observatory before Halley took over. She was the granddaughter of Ralph Cooke, who was rector at Burstow from 1637 till Flamsteed succeeded him in 1684, and she is buried with him in the church.

The location of the church is signposted on the B2037, which branches off A264 from East Grinstead towards Horley. The church is normally open in the daytime; the rectory is next door.

CANTERBURY (Kent)

When you make your pilgrimage to the celebrated cathedral, don't fail to visit the King's School, lying within the cathedral precincts on the north side, one of the oldest schools in Britain. Thomas Linacre was an early scientific student here around 1475, while the school was still a monastic institution. Henry VIII established the present secular grammar school in 1541 after he had confiscated the monastic properties; Richard Boyle, the father of Robert Boyle, and William Harvey were among the subsequently famous boys in the 1580s. The school has always been (and still is) a most rigorous school, its statutes calling for expulsion of any boy "remarkable for extraordinary slowness and dullness or for a disposition repugnant to learning" in order that "he may not like a drone devour the honey of the bees". Is this where the often crude and rapacious Earl of Cork (Richard Boyle) received the germ of inspiration for better things, resulting in his sending his sons Robert and Francis to Europe for an expensive and rigorous education?

Today's school retains the high standards of those days, but of course admits girls as well as boys. It has 700 pupils and annual fees of around £10 000 per pupil. One difference from the old days is the absence of choristers, members of the cathedral choir, subject to less stringent educational requirements than the rest. They actually persisted until after World War II, when a separate choir school was created for them.

Note: For foreign visitors King's School may be a particularly convenient way to see an English "public school" in action. Access is easy — one just continues north beyond the infirmary cloisters in the cathedral precincts. The school has recently expanded across Broad Street into what used to be a training college for monks and the stream of students crossing the street from one class to the next provides an alternative signpost to the main part of the campus. Note that the pupils wear uniforms, including the obligatory wing collar for boys. Purple robes designate prefects, the peer group guardians of law and

order at all such institutions. Classes are held mostly in the mornings, afternoons being devoted to other activities: organised sports on some days and socially relevant projects on others — the latter a truly modern touch.

DARTFORD (Kent)

Here we have an example (relatively rare in the technology field) of a brilliant inventor, who in his lifetime was denied his due and died a pauper. Richard Trevithick, designer of the Cornish engine (*see* CAMBORNE & REDRUTH, Cornwall p.123), and George Stephenson, builder of the famous "Rocket", are nowadays given comparable credit for converting steam power into railway motion. But whereas Stephenson ended his days with honour and prosperity in Derbyshire, Trevithick, at the time of his death in 1833, was working at a menial job at Hall's engineering works here in Dartford, and was carried to his grave (an unmarked pauper's grave in St. Edmunds churchyard) by his workmates. The parish has tried to make amends by a memorial tablet in the south aisle of Holy Trinity Church, paying tribute to "the memory of one whose splendid gifts shed lustre on this town, although he was not permitted to enjoy the fruits of his labour here."

Dartford is close to the Thames crossing of the M25 motorway; the church is in the town centre, close to the railway station. It is open on Thursday and Saturday mornings; other days only for services. Phone (01322)-222782 (Vicarage), (01322)-343243 (Tourist information).

DOWNE (near junction 4 of M25 motorway, Kent)

Charles Darwin's search for a suitable home, as described in his autobiography, strikes a familiar note: "After several fruitless searches in Surrey and elsewhere, we found this house and purchased it." The house was Down House, a secluded property at the edge of the village of Downe, and it was to remain Darwin's home from 1842 until his death in 1882 — almost the entire body of his work originated here. The preservation of the house as a national memorial is due to the benevolence of a London surgeon, Sir George Buckston Browne, who stepped in to prevent its sale in 1927. It is now partially restored as it was in Darwin's lifetime and its spacious grounds, too, are being maintained, so that the visitor can walk in the same gardens that served as rest and inspiration for the originator of the theory of evolution. (The house itself was and still is rather ordinary in external features. Darwin added the bay windows in the back as soon as he moved in and the verandah in 1874.)

The ground floor of Down House serves as a museum, and recently has become a satellite of the Natural History Museum in London. The drawing room and the "Old Study", are furnished more or less exactly as they were when Darwin lived here. The study was his daily workplace for 35 years and we can see the chair by the window where he did his writing, with a cloth-covered lapboard on which to rest the paper; his microscope stands close by. The study is in the front of the house and the windows face the public lane which runs alongside — Darwin had the road level lowered and put in the wall we now see in order to gain more privacy. Even then passsers-by on horseback could look in.

The most interesting part of the visit is the Charles Darwin Room, formerly the dining room, now a display room for paintings, photographs and showcases of memorabilia. It contains a model of _H.M.S. Beagle_, for example. Another room is devoted to manuscripts and books of grandfather Erasmus Darwin and the "New Study" (Darwin's workplace after 1877) houses a rather superficial exhibit on evolution, from when the earth was "a mass of whirling gases" until the appearance of modern humans.

The upper floors are presently occupied by private tenants, but there are plans afoot to convert the entire building to a Darwin study centre.

Down House is open from March to 15 December, Wed–Sun, 1–5. Phone (01689)-859119.

Down House, where Darwin wrote The Origin of Species.

EAST TYTHERLEY (near Romsey, Hampshire)

For 17 years, from 1846 to 1863, there was an important Quaker school near here, Queenwood College, one of the first where pupils did practical work instead of relying solely on lectures and demonstrations. Two of its earliest teachers were the chemist Edward Frankland (*see* LANCASTER, Lancashire p.228) and the many-talented John Tyndall (*see* HASLEMERE, p.108) — they became friends here and left together in 1848 to study in Germany, where knowledge of chemistry at the time was far more advanced than in Britain. The old school has been demolished, but Queenwood Farm, north of the village, occupies the site. The present fine residence on the property is the former headmaster's house.

Queenwood Farm is on an unnumbered road, leading from Broughton (B3084) south to East Tytherley.

FOLKESTONE (Kent)

William Harvey, born in Folkestone in 1578, is celebrated by a fine statue on The Leas, Folkestone's grand upper esplanade, parallel to the sea front on top of the cliffs. The church of St. Mary and St. Eanswith, tucked away into a quiet area at the east end of The Leas, has a Harvey Aisle and a window in his honour. It is interesting that Folkestone, though an important port and centre of commerce, did not have any regular school at the time. Harvey presumably received the rudiments of an education at home or church, for he could not otherwise have qualified for entry, at age 9 or 10, to King's School in Canterbury (*see* p.103).

The church is normally open during the day. Phone (vicarage) (01303)-252947.

GOSPORT (near Portsmouth, Hampshire)

Haslar Royal Naval Hospital in Gosport is a historic place. James Lind (1716–1794), who was chief of the hospital from 1758 until he died, is considered to be the founder of naval medicine, and is best known for his convincing demonstration of the efficacy of citrus fruit for the prevention of scurvy on long ocean voyages. James Cook's experiences on his voyages of exploration (*see* WHITBY, North Yorkshire p.251) indicated something of the sort, but Cook himself tended to think (falsely) that sweet-wort, an alcoholic malt beverage, was just as good.

A century later, Thomas Huxley's career began here. He had a medical degree, joined the Navy and was assigned for service at Haslar. He was reassigned to *H.M.S. Rattlesnake* in the capacity of ship's doctor — his interest in zoology developed during the 4-year voyage of the ship.

Plans are afoot as this is being written to use Haslar for training of medical personnel from all branches of the military, in which case its naval origins may become blurred. In any case, the site is necessarily under heavy security and visitors can do no more than peek in from the outside.

The hospital has a small museum, but it is open only every other Wednesday, 2–4. Phone: (01705)-584255, ext 2603.

HASLEMERE and HINDHEAD (Surrey)

Haslemere was the home (late in life) of one of the more colourful characters of 19th century science, John Tyndall (1820–1893). He was born in Ireland, came to England to work, was often in trouble because of his activities as a social protester, felt chronically threatened by poverty and always sought more work than he could handle to increase his income. He was fortunate to earn the respect of Michael Faraday and became professor at the Royal Insitution and eventually Faraday's successor as superintendent. He was almost as successful as Faraday as a lecturer (e.g., he gave the Christmas lectures 12 times), but his research was scattered over a wide range of subjects and seemingly always controversial. He is best known for his work on what has become known as the Tyndall effect, the opacity due to scattering of light by large particles in air, but he also wrote books on glaciers, mountain climbing, diamagnetism, etc., many of which were popular and earned him extra money, but did not sit well with his scientific colleagues. In 1876 he married a wealthy lady and thereby finally achieved financial security and a fine home here in fashionable Surrey. His wife was 25 years his junior and absolutely devoted to him, but even this serene part of his life had a lurid end, for Tyndall became seriously ill and bedridden and he died from an accidental overdose of chloral, tragically administered by his wife herself. (She went on to live to age 95, dying in 1940.)

The Tyndall house was in Hindhead and still stands, unchanged in external appearance, but now divided into flats and part of a small estate bearing the Tyndall name. Across the road is Tyndall's Wood, land given by Mrs. Tyndall in her will to the National Trust, part of Hindhead Common and adjacent to the Devil's Punch Bowl, of which one gets a spectacular view from the A3. Tyndall is buried in the graveyard of St. Bartholomew's Church in Haslemere.

Haslemere Museum. Tyndall's choice of this area for his residence was based on scientific research — he sampled and analysed the air in many places around London and found this the most salubrious. This recommendation led several other scientists to live in this area when they retired, among them the chemist A.W. Williamson (_see_ LONDON p.75), the geologist Archibald Geikie, and Charles Darwin's controversial cousin, Francis Galton (_see_ LONDON p.84). The small Haslemere museum, at the north end of the town's High Street, is a good source of information. They are in the process of developing Geikie archives, for example, to include documents, letters and sketches — Geikie was a skilful artist.

The Tyndall estate and Tyndall's Wood are on the A287 (Hindhead to Haslemere), just a few yards south of its inter-section with the major trunk road the A3. The museum in Haslemere is open all year, Tue–Sat, 10–5 (summer) or 10–4 (winter). Phone (01428)-642112.

ISLE OF WIGHT

The Isle of Wight illustrates a maxim that applies (probably uniquely) to English science. No matter where you may go for sightseeing or recreation, no matter how seemingly remote from present centres of commerce or transport, you will almost inevitably find that the place has had an unex-pected role in the history of science. In addition to the tourist attraction that may have led you there (the Cowes sailing races, for example, in the Isle of Wight, or the house where Queen Victoria died) less-publicised sites of con-siderable interest may lie just around the corner.

Freshwater, near the western tip of the Isle of Wight, is of interest to us because it is the birthplace of Robert Hooke. His father was a curate at All Saints Church, in the heart of old Freshwater village, by the saltings of the river Yar. The lo-cation is lovely, hardly changed at least in recent centuries, at the top of a hill now called Hooke Hill. At the foot of the hill is a stone memorial, with the message: "Robert Hooke. Born nearby 1635. Physicist, scientist, architect and inven-tor." Many ancient cottages line the road up the hill and one of them was probably the Hooke family home, but it is not known which one — Robert's father was only a curate, an underling to the rector. (Did Robert remain conscious of social inferiority all through his life? Was it the cause of his notorious belligerence towards Newton and others?)

Freshwater also enjoys some note for having been the principal home of the poet Lord Tennyson after he became poet laureate. He built Farringford, a magnificent mansion, for his residence. It is now a plush hotel, with its own 9-hole golf course and other amenities; the room that used to be Tennyson's library has been kept intact and is essentially a small museum. Even this poetic shrine has connection with the history of science, for Tennyson was of the Victorian period when scientists and literati mixed to some extent. Lord Kelvin was a particular friend, one of twelve pall-bearers when Tennyson was buried in Westminster Abbey in 1892. The Kelvins in fact continued to be friends of the family after the poet's death; Lord and Lady Kelvin were visitors at Farringford in June 1898 when they were taken to see the nearby Marconi station for wireless telegraphy (*see* below). Kelvin, who had many years earlier been the prime mover of cable telegraphy, sent a wireless message to George Stokes in Cambridge, described in a famous letter in which he expressed his pleasure at being able to send this message "through the ether" instead of along wires. A chance remark like this sometimes reveals more than a page of text — denial of the existence of the ether would have been inconceivable to Kelvin and most of his colleagues, although the failed experiment to demonstrate its existence was by then a decade old. (Einstein eventually abolished the idea as "superfluous".)

The Needles. Every visitor to the Isle of Wight should take the walk to the Needles Headland on Alum Bay — an extension of the Tennyson trail from Freshwater if one wants to make a long excursion of it. It is where the Marconi transmitting station was located and it is arguably the most spectacular bit of scenery on the island. (Later on, when transatlantic telegraphy became the goal, Marconi's operation was moved to another beauty spot, this time to the Lizard peninsula in Cornwall.)

LEWES (East Sussex)

Gideon A. Mantell (1790–1852) is one of two famous sons of Lewes, the other being Thomas Paine, the advocate of popular revolution against oppressive governments, said to have been influential in the advent of the American revolution. Mantell was a practising physician and an amateur palaeontologist, and is best known for his discovery of the first dinosaur (*terrestial* as distinct from *aquatic*) ever to be described properly — a "momentous event", in the words of his biographer. Mantell named his find *iguanadon* because its teeth resembled the teeth of the modern lizard iguana, but this turned out to be misleading because no actual relationship exists.

Mantell was a prolific writer, his major scientific work being *The Fossils of the South Downs* published in 1822. However, he also wrote books intended to popularise his field among the lay public, e.g., *The Wonders of Geology*, *Medals of Creation* and *Thoughts on a Pebble*.

The house where Mantell was born and lived for over 40 years is Castle Place, at 167 High Street, between St. Michael's Church and Lewes Castle — it is marked by a plaque. The Barbican House, practically next door at the castle gates, is a general museum for the Sussex Archaeological Society — its library contains diaries and

Plaque outside the house in Lewes where Gideon Mantell was born and lived for 40 years.

other documents related to Mantell, which are available for research to anyone interested.

Barbican House is open all year, Mon–Sat 10–5.30, Sun 11–5.30. Phone (01273)-486290.

Tilgate Forest. Tilgate Forest, where most of Mantell's local finds were made, is south of Crawley, almost 20 miles from Lewes. Some of the area is now urbanised and even the remaining green parts are sliced in half by the M23 motorway. "Tilgate Forest Park" (north of the motorway) is a family recreation area with a nice lake and an adjacent golf course. A small area of relatively undeveloped forest, traversed by a couple of footpaths, can be found south of the motorway.

Tilgate Park is reached by taking the A23 north from the present terminus of the M23 (junction 11), followed by a signposted right turn. The undeveloped area is just to the east of Pease Pottage, a village signposted at junction 11.

OCKHAM (near Guildford, Surrey)

"Occam's Razor" is a vital maxim for students of at least the physical sciences. It's a warning to beware of too fanciful hypotheses — if one novel idea, plucked seemingly from nowhere, will account for something previously unexplained, it is worth a follow-up, but if it takes more than one, beware! The rule is named after William of Occam (*ca* 1285–1347) — "What can be done with fewer assumptions, is done in vain with more" is the way he put it. William was born in this tiny Surrey village and undoubtedly attended All Saints' Church (which dates from 1160), for he could not have gone to Oxford, as he did, without the priest's help. It was in Oxford that he advocated his maxim. It may well have had unpopular theological implications; in any event William was summoned to the papal court in Avignon to answer charges of heresy and soon became an active participant in the great schism of the Catholic church, between the Pope in Avignon and the anti-Pope in Rome. He died in Munich, where there is a memorial and an Occamstrasse named after him. All Saints' Church has put in a memorial window for him, where it describes him as "Doctor Invincibilis" (and where "Occam" is still spelled thus, with a double "c").

There is an interesting postscript. The fictional hero of the recent popular novel, *The Name of the Rose*, by Umberto Eco (which takes place at the time of the schism) describes himself as a friend and protégé of William of Occam and of Roger Bacon as well, but his journeys and philosophical discussions bear an uncanny resemblence to what we know about Occam himself. William of Baskerville, as Eco calls his chief character, solves the mystery of serial murders in an Italian Benedictine Abbey by applying the "philosophical razor" in truly brilliant style. We recommend the book to get a feel for the times — especially the jealousy with which access to the library was guarded when all its books, containing herbal remedies and other secrets of what we might now think of as within the province of science, had been laboriously copied by hand.

If Ockham is approached from the west (Woking, Ripley), a signpost to the church is found on the right at Guileshill Lane as soon as one enters the village. The church has a tourist attraction in the form of a 7-lancet window which dates back to the medieval origins of the church. The church tends to be closed outside of services, but keys can be obtained from the Hautboy hotel in the village. Phone (rectory) (01483)-225358.

PILTDOWN (near Uckfield, East Sussex)

The sites of anthropological discoveries are sometimes huge gashes in the earth, created by quarrying or road construction, or spectacular deep caves, their entrances long hidden from sight. The Piltdown site at Barkham Manor is more prosaic, a place in a field near a ditch, where there used to be a gravel pit in which amateur diggers had found bones and tools in the past. It is part of what used to be a diversified manorial estate, but has recently been converted into an ambitious 26-acre vineyard, complete with modern wine-making equipment, and a barn and shop where refreshments can be purchased and the wines can be tasted. White wines called *Barkham Manor* and *Piltdown Man* are among the products. A pub called "Piltdown Man" stands near the point where the side road to the vineyard leads off from the A272.

A self-guiding trail (explanatory leaflet provided) takes visitors through fields planted with different grape varieties and through the winery buildings. The place where the skull was found is one of the numbered spots on this trail, marked by two monuments. One, a proud stone monument, was obviously erected before the hoax was discovered. The inscription reads:

> Here in the old river gravel Mr. Charles Dawson, F.S.A., found the fossil skull of Piltdown Man 1912–1913. The discovery was described by Mr. Charles Dawson and Sir Arthur Smith Woodward in the Quarterly Journal of the Geological Society 1913–1915.

An adjacent plaque, more factual and put up many years later, explains that the remains were a deliberate hoax. It states that the first announcement came 18th December 1912, identifies Dawson as a local solicitor and amateur archaeologist and Woodward as "keeper of geology" at the British Museum. It also tells us that Woodward and Dawson returned to the site in 1913 with the French Jesuit, Teilhard de Chardin. The latter, digging around some more, uncovered a tooth, subsequently identified as of canine origin. The proof of the hoax is correctly ascribed to Dr. K.P. Oakley.

The vineyard and buildings are open to the public daily except Mondays. Phone (01825)-722103.

PULBOROUGH (near Midhurst, West Sussex)

The attraction here is Parham Park, an Elizabethan house and garden, filled with portraits and antique furnishings, one of the popular showplaces of the area. It is also the sole repository of mementoes of Joseph Banks (1745–1820), the scientist who accompanied James Cook on his first voyage of discovery and subsequently became a leading figure in British science for many decades (*see* REVESBY, Lincolnshire p.212) — though the association of the house with Banks is quite indirect. What happened is that Banks died without direct heirs and left his effects to a sister, from whom they eventually passed to the wife of the owner who bought Parham House for his residence in 1922. The house has been turned into what is effectively a museum and one room (called the Green Room) is set aside to exhibit the Banks collection.

The most interesting items are two paintings by the artist George Stubbs, one of a kangaroo and one of a dingo dog, based on skins brought back by Banks, inflated to provide a not altogether accurate representation of the whole animals. They put us in the mood of the times, however, and there is no other place where we can get this kind of retrospective insight into the sensation caused by exotic new species brought back from Australia. The room also contains Banks's globe, specimen chest and several portraits.

Parham House is well signposted on the A283 between Pulborough and Storrington. It is open to visitors April to end September, Wed, Thu, Sun and Bank Holiday Mon, 2–6. Phone (01903)-744888.

SELBORNE (near Alton, Hampshire)

The systematisation of flora and fauna by Linnaeus and his followers triggered a popular interest in "Nature" in Britain in the second half of the 18th century, which seems never to have subsided. It was then that books about the countryside began to be avidly read and the most influential of them was _The Natural History of Selborne_, a literary compilation of notes and letters by Gilbert White (1720–1793), local curate and unflagging observer of the plants and animals around him.

The Gilbert White Museum in Selborne occupies the house in which White lived almost all of his life and it admirably recreates the atmosphere of the times. Although the house was extended by later owners, the original layout and size are made apparent via the guide book. One entire room is devoted to the creation of "the Book" and includes a good historical guide to the man and his work. The gardens are in the process of being restored in part as they were in White's day. There is a collection of family portraits but (strangely) none of the man himself are extant. St. Mary's Church, across the street from the museum, where White's grandfather (also named Gilbert) had been the vicar for nearly 50 years, contains several memorials — the most eye-catching is a window, illustrated by various plants and animals, inscribed to White, whom it calls a "humble student of nature". White is buried in the churchyard and his gravestone carries (as he wished it) only a simple inscription: "G.W. 26th June 1793".

The land behind the museum rises steeply — "a long hanging wood called the Hanger," is how White himself describes it. It is still called the Hanger and is part of a National Trust countryside area, with many footpaths, one of them a steep zig-zag path to the top of the land that was laid out by Gilbert White himself and one of his brothers.

Selborne is south of Alton on the B3006. A single car park on the village High Street serves visitors to both the museum and the National Trust area. The museum is open daily mid-March to end of October, weekends only the rest of the year, 11–5.30. Phone (01420)-511275.

Gilbert White Museum in Selborne. The "Hanger", where he studied plants and animals, rises steeply behind it.

116

3
The Southwest

AVEBURY (near Marlborough, Wiltshire)

We find here the largest prehistoric earthwork and stone circle in Britain, so large that much of the village of Avebury and the highway that runs through it lie within the circle. Unlike Stonehenge, this is not an isolated monument but part of an area rich with other remains, attesting to continuous occupation from 3500 B.C., much earlier than the circle itself. Peripheral places of interest include a stone-lined "avenue", leading to a burial site a mile away. A museum contains pottery, tools, stone axes, etc., and explains the local history. We are told that the people who lived here did not yet have the use of the wheel — which makes their grand construction projects all the more remarkable. (Avebury was first recognised for what it is by John Aubrey, one of the earliest fellows of the Royal Society, who surveyed the area for King Charles II. Aubrey is best known for his book *Brief Lives*, which contains crisp character sketches of several of his Royal Society colleagues.)

The site is visitor-friendly, with well-marked footpaths, large car parks and pubs for food and drink. Everything is open all year.

BATH (Avon)

Bath was a spa resort at the time of the Roman occupation
of England and has ever since remained a fashionable place
to live or visit. The celebrated Royal Crescent (built 1767–
1774) is one of the most graceful and beautiful examples of
domestic architecture in all of England. Bath's perhaps
unexpected association with science stems from the presence here of William and Caroline Herschel, who actually
came to Bath as musicians, William as organist of the
fashionable Octagon Chapel and Caroline as a vocalist. But
they started to build telescopes and map the skies and eventually acquired fame and royal patronage as astronomers —
they left in 1782 at the king's command to take up residence
and set up their telescopes closer to Windsor Castle (*see*
SLOUGH, Berkshire p.163).

Herschel Museum. Herschel's house at 19 New King
Street, where he lived and worked from 1777–1782 (and
where the planet Uranus was discovered) is now a museum,
located close to the city's main tourist sights. The telescopes, built with the help of another brother, Alexander,
were the reflecting kind, with about the best resolving
power of any instruments of the day. William cast the mirrors himself, using the special alloy *speculum*, and
Alexander built the bodies out of wood. Brightly varnished,
they are beautiful to behold — there is a replica in the
museum on the ground floor. The work rooms where the
telescopes were built are in the basement of the museum and
contain old tools and lenses. Upstairs is a reconstructed
music room, furnished to give an idea of what it might have
looked like in Herschel's day. It has an old pianoforte and
the organ keyboard rescued from the old Octagon Chapel
when it was converted to other use. (The former chapel itself
is on Milsom Street, now a lecture room that is part of the
National Centre of Photography.)

Visitors may be intrigued by a picture on the wall in the
museum, showing practically everybody in the world of science around 1800, supposedly assembled together for discussion at the Royal Institution in London — Herschel is
there, of course, together with Cavendish, Dalton, Jenner,

Rumford, Watt, and many others. This picture actually represents an imaginary scene, painted in 1862 — no such assembly ever took place.

The museum is open all year, daily 2–5. Phone (01225)-336228.

Bath Abbey.   Bath Abbey is the final resting place of Robert Malthus (1766–1834), political economist and author of the famous essay on population, setting out the principle that unchecked human population growth would eventually outstrip our food supply. There is a memorial on the porch, to the left as one enters the church, with an eloquent inscription, said to have been written by William Otter, Bishop of Chichester, who was a great friend and a frequent travel companion. "One of the best men and truest philosophers of any age or country," the inscription reads, an evaluation that is probably not shared by most of today's social scientists, who tend to regard Malthus's proposed solutions for the population problem as illiberal. Charles Darwin read Malthus's essay before he published his *Origin of Species* but the main ideas of his theory had certainly been established in his mind long before then; a direct influence of Malthus on A.R. Wallace may be a little more plausible.

Malthus's home for the last years of his life, the place where he died, is at 17 Portland Place, in an area of fashionable homes with good views.

BRISTOL (Avon)

Isambard Kingdom Brunel.   Bristol's most celebrated citizen is the construction engineer Isambard Kingdom Brunel (1806–1859), master of the use of iron, who built the railway that joins Bristol to London and designed the graceful suspension bridge across the Avon Gorge from Clifton Village. Brunel's greatest achievement was the building of ocean-going steamships and one of them, the *S.S. Great Britain*, is in the process of reconstruction in the dock where she was actually built. She was the first large steamship to use propulsion by means of a screw propeller (in place of paddles), the feasibility of which had first been demonstrated by Brunel in experiments with a smaller ship, appropriately named *Archimedes* — it is worthwhile to remember that the physical principle involved in screw propulsion to move an object through the water is exactly the same principle that Archimedes used more than 2000 years earlier for moving water uphill against the gravitational attraction of the earth. (Brunel, unlike many earlier giants of engineering, was in fact well educated in physics and mathematics, having been sent by his father to Paris to learn his trade. The father, Marc Isambard Brunel, was as famous as his son for his giant construction marvels, but he lacked his son's theoretical background. The father designed and built the world's first underwater tunnel, across the Thames at Rotherhithe, on the east side of London, but many disasters occurred during the building — the son almost lost his life trying to save others on one occasion when water flooded the tunnel.)

The suspension bridge and the *S.S. Great Britain* are part of the standard Bristol tourist itinerary. Bristol has a gadgety "Exploratory Science Centre" of the kind now seen in many cities — it is at Temple Meads, on the east side of the city, in the original engine shed set up by Brunel in 1840 for the Great Western Railway. Both this museum and the *S.S. Great Britain* are open 10–5, seven days a week.

Beddoes Pneumatic Institution. Thomas Beddoes (1760–1808) moved to Bristol from the Midlands, where he had been a member of the Birmingham Lunar Society. He had been impressed there by the new "airs" discovered by Joseph Priestley and convinced that they might be beneficial for curing tuberculosis and other respiratory diseases. He founded his Pneumatic Institution in Clifton by subscription in 1798 to test the matter and, seeking a research person to investigate the chemical aspects, found young Humphry Davy (then aged 19!), apprenticed at the time to an apothecary-surgeon. Davy was lucky to come out alive from some of the foolhardy experiments he did, but gained fame with his discovery that nitrous oxide ("laughing gas") could induce a jolly state of mind — it became a popular adjunct for adventurous party-givers. But no true therapeutic benefits were found for any of the gases and Beddoes' Institution eventually became just a plain hospital. Humphry Davy, on the other hand, went on to a truly momentous career, not just in terms of his own discoveries, but equally for his influence on the course of the history of British science. It is dubious whether the Royal Institution in London would have survived its early days without him, and it is virtually certain that Michael Faraday would never have come close to a career in science without his appointment as Davy's assistant there. It is truly legitimate here to ask how different our world might have been without Thomas Beddoes' faith in "pneumatics".

The building formerly occupied by the Pneumatic Institution is at 6 Dowry Square in Clifton and is marked by a plaque. Dowry Square is just a few steps from Hotwell Road (via Dowry Parade), about halfway between the *S.S. Great Britain* and the Clifton suspension bridge. Access is made a trifle complicated by major motor roads that enter the city close to this point.

BRIXHAM (Torbay, Devon)

Brixham has the distinction of being the place in England where the great antiquity of the human species passed a watershed, from conjecture to certainty. Excavations from a cave here in 1858 provided *incontestable* evidence for the co-existence of tool-making primitive men and women and Pleistocene mammals, extinct for a million years. The remains have been dispersed to national museums, but the cave is still on the map, its entrance situated beneath a row of terrace houses on a hilltop road. It was open to the public until a few years ago, but has since been declared unsafe and boarded up. Space has been left at the top of the boards to allow us to look down into the cave from street level.

The cave entrance is at 107 Mt Pleasant Road. It is close to the town centre, at the end of Cavern Road (for pedestrians only), which rises steeply from Bolton Road.

CALNE (near Chippenham, Wiltshire)

The attraction here is Bowood House, where Joseph Priestley discovered oxygen (*see* CHEMISTRY p.42). Priestley was in the service of the Earl of Shelburne from 1773 to 1780, formally as librarian, but actually as an aide of broader usefulness, a sort of resident intellectual. Bowood House was (and still is) the country home of the Shelburnes and Priestley used to spend his summers here and his winters at the Earl's London house. Shelburne encouraged and financed Priestley's scientific work — the studies of gases (summarised in the famous book *Experiments and Observations on Different Kinds of Air*) were mostly done during this period, the discovery of oxygen being made on 1 August 1774, in a small room set aside at Bowood specifically as a laboratory for Priestley. It is difficult to imagine more sumptious quarters for laboratory research!

A leaflet for visitors explains that the estate has "been added to or reduced, reflecting social changes and family fortunes, like the continual movement of the tides." In fact, what used to be the "big house" was demolished in 1955. The orangery and its attendant pavilions remain and they contain the Shelburne library, Priestley's laboratory next door to it, a chapel, a sculpture gallery, a picture gallery, and plenty of space for the family residence and servant quarters. The laboratory has none of Priestley's equipment, but it contains original letters from Priestley to the Earl about

Bowood House, where Priestley discovered oxygen.

122

Priestley's scientific work and his supervision of the education of the Earl's children. The adjacent library is decorated by sixteen huge black basalt vases with "Etruscan" designs, which were made for the family in 1813 by Josiah Wedgwood II, but represent a style of vase popularised much earlier by the first Josiah Wedgwood. The latter provided Priestley with much of his scientific ceramic ware — presumably lacking "Etruscan" adornments. (_See_ STOKE-ON-TRENT, Staffordshire p.186)

The entrance to Bowood House is off the A4, midway between Calne and Chippenham. The grounds are elegantly landscaped and include a lake, with a Doric temple on its banks, terraced gardens leading up to the main house entrance, an arboretum and much more — plus a commercial garden centre, an adventure playground for children, and other ventures designed to aid the family fortunes. The site is open April to end of October, daily, 11–6. Phone (01249)-812102.

CAMBORNE and REDRUTH (Cornwall)

Richard Trevithick (1771–1833), nowadays considered to have equal claim with George Stephenson for the invention of the steam locomotive, was born in this community bursting with mining landmarks, and lived here till 1815. Specifically, he built a revolutionary _high pressure_ steam engine which represented a huge advance over the Boulton–Watt engines and came into wide use for raising ore, refuse and men from the deep Cornwall tin mines and elsewhere. Right from the beginning he also visualised the use of this engine for locomotion and built several successful models, but this product of his inventive mind found no ready acceptance because of the expense involved in building tracks that could carry mobile engines. Horses had surer footing on existing paths and continued as the _mobile power_ of Cornish mines until the price of horse fodder became prohibitively high. Trevithick thus never gained the national prominence that Watt or George Stephenson had. He spent his resources with reckless abandon on ever newer inventions and died a pauper in Dartford, Kent.

Trevithick's birthplace is marked by a stone in East Pool, opposite an old entrance to the only remaining commercial tin mine in Cornwall (South Crofty) — there were hundreds of them in the boom years of the 1860s. The cottage with thatched roof, where Trevithick lived from 1810 to 1815, is in Penponds, just west of Camborne, still used as a residence and marked by a plaque. (There is also a rather poignant memorial in Dartford, where he died.)

Cornish Engines. The National Trust has preserved two Trevithick-type beam engines (both dating from around 1890) at Pool, half-way between Camborne and Redruth; one is a pumping engine for clearing mines of water, the other a winding engine for carrying men and ore. They are huge in size, housed in buildings to match — an 8-foot-thick wall, for example, to support the beam of one of the engines — and one of them is in perfect working order, authentic except that power to the machinery is supplied by electricity rather than steam. Attending staff are either engineers or former miners and expert in explaining both the engines and their use in the mines. An added bonus is that

East Pool Whim, a Cornish engine preserved by the National Trust.

many of the guides are very knowledgeable about the local geology — a single question can elicit a lengthy lecture on the properties of granite, the rate at which ground water passes through fissures in the strata, chemicals to be found in the water from flooded mines, etc.

The Cornish Engines sites stand on opposite sides of the A3047 and are signposted. They are open from the end of March to the end of October, daily 11–5.30. Phone (01209)-216657. The National Trust also displays a beam engine that has been restored to use steam as the actual source of power, but it is at Pendeed, near St. Just, about 20 miles west of Camborne — see PENZANCE. (Use same phone for information.)

DARTMOUTH (Devon)

The age of steam — when we use the phrase we tend to think first of James Watt, who turned the world around by improving the Newcomen engine and combining with Matthew Boulton in Birmingham to produce it in quantity. But what if there had been no Newcomen engine to improve? Thomas Newcomen (1664–1729), native son of Dartmouth, was in fact the real pioneer; for more than 50 years his steam engine was the only one available, the only source of mechanical power that did not depend on animals or rarely available water power. Its most important use was to raise water from underground mines, allowing continued production when shallow surface deposits of coal or minerals had been exhausted.

A genuine Newcomen engine has been installed in an engine house on the Dartmouth water front and electrical simulation of its operation can be turned on if requested. More important, a poster display provides beautifully clear explanation of the principles involved, based on properties of air, steam and water. Later contributions by Smeaton and Watt are defined; good diagrams of working engines are particularly valuable for non-engineers. We are told that the thermal efficiency of the Newcomen engine was less than 1% and Watt's later 3–4% was clearly an enormous improvement. The only omission in the display is insufficient emphasis on the theoretical limit on attainable efficiency (the basis for the science of thermodynamics) —

Newcomen's and Watt's figures are not nearly as bad as they may at first appear to be.

The life of Newcomen is also covered briefly and emphasises his lack of formal training; he was an ironmonger by trade and collaborated with John Calley, a plumber, to build his first engine. The house on the quay where Newcomen lived is marked by a plaque.

The Newcomen engine house is attached to the beachfront tourist information office. It is open all year, daily 9.30–5.30 in summer, Mon–Sat 9.30–4 in winter. (The approach to Dartmouth by ferry from the Torbay area is both quick and panoramic — highly recommended.)

HIGH LITTLETON (near Bath, Avon)

Rugbourne Farm, on the outskirts of this village, is the "Birthplace of English Geology", so named by William

The "Birthplace of English Geology". William Smith took lodgings at this farm in 1791 while on a surveying job, and it was here that he resolved to make his map of geological strata for all of England and Wales.

Smith (1769–1839) himself, the presumed father. Smith was sent to this area by Edward Webb, the Gloucestershire surveyor for whom he worked, in order to survey and value a large estate. Smith took lodgings at Rugbourne Farm and, as he worked on his assignment and took on other local projects, his mind was repeatedly drawn to the succession of geological strata wherever digging took place — mines, quarries and the Somerset Coal Canal, in the construction of which Smith was personally involved. It was at High Littleton that he discerned the need for a definitive map of these strata throughout England and Wales and resolved to fill the need himself. It was to be 1815 before the map was completed — all of England and Wales and part of Scotland on a scale of 5 miles to the inch. The maps were coloured, with shading to indicate how the layers were superimposed on one another. About 400 copies in all were issued, fewer than 100 have survived.

Rugbourne Farm still exists, a relatively modest operation by its appearance. The farmhouse still stands, the exterior virtually unchanged since Smith lived here. Lodgers are no longer taken. There is no plaque at the house, but there is a stone with an inscription at the place where the lane to the farm turns off the paved road — it was mostly hidden by weeds when we were there.

High Littleton is on the A39, south of Bath. The farm is off an unnumbered road at the south end of the village (signposted Timsbury); there is a caravan park on the lane leading to the farm — the inscribed stone is at the turn-off.

HORTON (near Chipping Sodbury, Avon)

William Prout (1785–1850) is a familiar name in the history of the atomic theory, not because of any spectacular discovery, but because he formulated a hypothesis (in 1815) that epitomises the central problem in relating chemistry to physics, the explanation for the distinction between the atoms of the different elements. He proposed that all atomic weights are integral multiples of the atomic weight of hydrogen; i.e., that hydrogen might be the primary matter from which all other elements are formed — it is so simple as to be almost in the category of wishful thinking. "Prout's hypothesis" was used for over a century as a catchword to define the basic problem, while the reality of an actual integral relation had its ups and downs. It triumphed in the end: we now know that every atomic nucleus is composed of an integral number of protons and neutrons and that atomic weights are never precisely integral because of a small dependence of measurable "mass" on cohesive energy, a manifestation of Einstein's $E=mc^2$.

Prout was the son of a tenant farmer on the Horton estate, but apparently held legal right to the tenancy he acquired by inheritance, for he was able to "give it away" to his brother. Prout had an M.D. degree and his scientific work was broad. He demonstrated in 1824 that gastric juice of animals contains hydrochloric acid, which at first seemed an impossibility — how can the stomach walls resist the high acidity?

Horton Court, the manor house of the Horton estate, is a National Trust property with a 12th century Norman hall. It is leased to tenants who live there, but open to the public Wed and Sat, 2–6, summers only. Neither the tenants nor the National Trust have any knowledge of Prout. We have not so far been able to pinpoint the precise location of the former Prout farm.

KINGTON ST. MICHAEL
(near Chippenham, Wiltshire)

John Aubrey (1625–1697), a not very serious antiquary and naturalist, is best known as a celebrity chaser, who wrote pithy and irreverent accounts of the people he encountered and what he learned from some of them. He wrote it all down in the form of notes, but dates and facts were often inaccurate or missing altogether. His *Brief Lives*, as found on most library shelves today, is an edited compilation, published in 1898. Its importance to us lies in the fact that scientists were among his biographees (Aubrey was one of the earliest fellows of the Royal Society) and his remembrances of them are sometimes the only personal information we have about them. Aubrey (born in Easton Piercey nearby) was an active member of the Kington parish. His curiosity about the Wiltshire countryside led him to discover what are now known as the "Aubrey Holes", surrounding the stone circles at Stonehenge — but at the same time he made the entirely unfounded suggestion that the monument used to be a Druid temple, a belief still held today by certain fringe groups. The parish church of St. Michael and All Angels has a memorial window to commemorate Aubrey and another local worthy, but the subject matter is biblical.

Kington St. Michael lies off the A429, north of Chippenham and close to exit 17 of the M4 motorway. The church is normally open in the daytime. Phone (rectory) (01249)-653839.

LACOCK ABBEY (near Chippenham, Wiltshire)

At Lacock we have a unique combination: the Fox Talbot Museum of Photography, dedicated to William Henry Fox Talbot (1800–1870), the inventor of photography in its present form, and Lacock Abbey, the Talbot family home, parts of which date back to the 13th century. The physical setting is lovely — even the adjacent old village of Lacock was largely Talbot property and is now maintained by the National Trust along with the Abbey. The museum itself is located in a former stable.

The museum admirably explains the principles of the photographic process and the history of its evolution. We learn that it began with the camera obscura, an ancient device used by artists to project an image of an external scene onto tracing paper, where its outlines can be traced with a pencil. "How charming it would be if it were possible to cause these natural images to imprint themselves directly and remain fixed upon the paper," said Fox Talbot, and set about doing that very thing. The Frenchman Louis Daguerre had the same idea, but he fixed his images on copper plates. Daguerre's results were at first more spectacular, but he could produce only a single picture per exposure. The Fox Talbot procedure led to images with light and dark reversed — for which John Herschel devised the term _negative_ — with the advantage that limitless positive copies could be made, once the problem of devising a durable base for the negative was solved. Many fascinating examples of early photographs by Fox Talbot and others are included among the exhibits.

_129_

Lacock Abbey is south of Chippenham, off the A350 and sign-posted. The museum is open end of March to end of October, daily 11–5.30. Phone (01249)-730459.

LAUNCESTON (Cornwall)

Here we have a story, fortunately rare and therefore very well known, of a lone scientist who did a truly original and important piece of work, but was denied credit and even publication by narrow-minded attitudes of authority — though in the end he did receive recognition and honour. The scientist was John Couch Adams (1819–1892), his feat was the virtual discovery of the eighth planet, Neptune, and the villain in the piece was none other than the Astronomer Royal, George Airy, who bore no malice towards Adams, but who held his mandate to be the making of precise observations for navigational purposes — original thought and scientific discovery were not on his agenda.

Adams was born on Lidcot Farm, which still exists about 7 miles west of Launceston, near the village of Laneast. He was a precociously brilliant mathematician, went to Cambridge for a degree and subsequently a fellowship. In 1843 (back at Lidcot for the long summer vacation) he devised new mathematical procedures to analyse apparent anomalies in the orbit of the seventh planet Uranus and demonstrated that they must arise from gravitational perturbation by a more distant planet, the approximate mass and orbit for which Adams was able to compute. He tried to persuade Airy to institute a search for the planet, to no avail. In the meantime (in 1846) the French astronomer Urbain Leverrier independently made similar calculations and in his case had not long to wait for confirmation — Neptune was observed close to its predicted position, and Leverrier was showered with honours. John Herschel, the son of William Herschel, the discoverer of Uranus, then took up Adams's claim for priority and Airy's telescopes found Neptune too when they finally looked for it. Adams eventually received his due, became director of the Cambridge Observatory and twice served as president of the Royal Astronomical Society.

Lawrence House in Launceston, owned by the National Trust but run as a local museum by the town council, has a bust of Adams and a couple of wall posters about him. Notebooks containing some of Adams's original calculations are said to be in the museum's possession, but could not be found on the day we visited.

Lawrence House is open April to early October, Mon–Fri, 10.30–4.30. Phone (01566)-773277 or 773047. (There is another tribute to Adams in the Cathedral city of Truro, *see* p.141.)

THE LIZARD (Cornwall)

The Lizard is England's most southerly point, made especially attractive by a spectacular segment of the Cornish Coast Path along the tops of the cliffs. Six miles northwest of the point itself, accessible from Poldhu Cove, is the Marconi memorial, a granite obelisk erected to mark the site of transmission of the first transatlantic wireless telegraph message. The message, consisting of the Morse code letter "S" (three dots), repeated at preset intervals so that it could be readily identified as a deliberate human signal, was sent out from here by staff of the Marconi Company in December 1901 — Marconi himself being at the receiving end in Newfoundland. On the other side of the Lizard, on the Coast Path around Bass Point, a plaque indicates the site of a building (the "signal station") where Marconi prepared and tested his equipment prior to the event.

How did Marconi, an Italian (and an avid patriot), happen to do his work in England? The ultimate reason is that his mother was Irish, a member of the Jameson family of Irish whiskey fame. When Marconi (initially inspired by Hertz's experiments with electromagnetic waves) was unable to interest the Italian government in the practical potential of his work, he appealed to his Irish cousins for help and one of them turned out to be influential in the London business world and able to find investors to finance Marconi's projects.

Goonhilly Downs. The Lizard peninsula is still today a

major radiotelecommunication centre — Goonhilly Downs, three or four miles inland from Lizard Point or Poldhu Cove, is a virtual forest of skyward-pointing dishes, beamed at satellites that handle millions of international telephone calls and television transmissions each year, to and from all parts of the world. The site has a visitor centre, which serves more as an advertisement for the telephone industry than as a source of scientific or historical information. Nevertheless, the displays and available literature do tell us about parabolic focussing and the usefulness of optical fibres, even if there is not much in the way of a physical explanation. And one can stand among the huge dishes and meditate in awe about the satellites, perched 22 000 miles above the earth, invisible to the eye, but unerringly targeted by the radio waves.

Goonhilly Earth Station is signposted on the B3293. It is open from Easter to the end of October, daily 10–6. Escorted bus tours operate frequently for access to the 7 largest dishes. Phone (01326)-221333.

LYME REGIS (Dorset)

The cliffs around Lyme Regis and nearby Charmouth have been a paradise for fossil seekers for two hundred years and remain so today. They are subject to a complex of rotational slides, sand runs, mud flows, and rock falls — land movements that continually alter the precise shape and character of the seaward edge and, in so doing, spew forth the fossilized remains of the life forms of millions of years ago. It was here in 1811 that a young girl, Mary Anning, discovered the first complete fossilised skeleton of one of the giant lizards — that of a 180 million year old marine lizard *Ichthyosaurus* — which she sold to the British Museum for £25. Mary Anning later became great friends with Oxford geologist William Buckland (who was a native of this region, born in Axminster, 5 miles from Lyme Regis) and found many more unique specimens — among them a new species of *Pterodactyl*. (For the uninitiated, which before our visit here included the present writers, it should be pointed out that the term *dinosaur* refers exclusively to ter-

restrial giant reptiles, _ichthyosaurs_ have a marine origin and
pterodactyls were flying species.)

Mary Anning's father (Richard Anning) was a carpenter
by trade, but he also sold fossils to Lyme Regis holiday
makers, who, just like today's visitors, loved to take home
souvenirs. Mary and her brother Joseph had gone beach-
combing for fossils even as children — they were
experienced collectors and immediately able to recognize a
new find when it appeared. They are buried in the graveyard
of the church of St. Michael the Archangel, with a single
tombstone for both. Mary is also commemorated by a
stained glass window in the church, erected in her memory
by the vicar of Lyme and by members of the Geological
Society of London, with an appropriate tribute inscribed in
the glass. There is a small museum not far from the church,
on the likely site of the house where the Anning family
lived. The focus is on fossils and on Mary Anning, and it
contains an _ichthyosaurus_ skeleton similar to that found in
1811. (Another fine _ichthyosaurus_ can be seen in the
Sedgwick museum in Cambridge. Adam Sedgwick bought
it from Mary Anning in 1835 for £50.)

Mary Anning's find occurred after a violent storm and, as

_The cliffs at Lyme Regis. Fossils still emerge today as the rocks
crumble._

we noted above, storms still cause rock to break away today and fossils can then be found by the diligent searcher. Large ammonites are quite common and even *ichthyosaurus* back-bone vertebrae still appear a few times each year. Caution is necessary, however. The rocks are steep and the beach at their base disappears at high tide, making it easy to be trapped without a way of escape.

It is worth noting that Henry de la Beche (1796–1855) is another "rock person" whose may have found his vocation at Lyme Regis. He lived here for a while as a youth, after having been dismissed in disgrace from military school. He was about the same age as Mary Anning and may well have known her, but he came from a quite different social stratum and was ambitious for national prominence. He was no beachcomber and his role in geology was on a broad scale, as a disputant in the "Devonian controversy" (*see* GEOLOGY p.35) and as founder of the Geological Survey.

The Lyme Regis museum is open daily from April to October, Mon–Sat 10.30–1 and 2.30–5, Sun 2.30–5 only. Phone (01297)-44 3370.

MILVERTON (near Taunton, Somerset)

Here we celebrate Thomas Young (1773–1829), who was born in Milverton, the son of a Somerset merchant and banker. He was one of the most versatile and brilliant men that English science has ever seen, but a dilettante, who himself later admitted that his work had often consisted of "acute suggestions", rather than fully documented airtight proofs. The spread of his interests is virtually incomprehensible in today's specialised world — he was a practising physician and published medical treatises; he effectively proved the wave theory of light (at the expense of Newton's corpuscular theory); he was the first to explain the phenomenon of surface tension at a liquid surface; he generated theories of elasticity and of ocean tides; he was the genius who worked out the key to the Rosetta stone, the first Egyptian hieroglyphics to be deciphered. But none of it was ever quite "finished". The credit for the *ultimate* decipherment of hieroglyphics, for example, is generally given to the

Frenchman, Jean Champollion, and not to Young. (*See* BRITISH MUSEUM in London p.74).

The house where Young was born and lived for some years is now called the Old Bank House. It is a substantial dwelling on North Street and carries an appropriate plaque.

Taunton. Taunton, the Somerset county town, has memorials to many Somerset worthies, Young included, in the Shire Hall, which dates from 1850. It is now used as a court house, and stands adjacent to the County Hall.

PAIGNTON (Torbay, Devon)

Oliver Heaviside (1850–1925) was a member of the so-called Maxwellians, who (between 1879 and, roughly, 1894) extended Maxwell's great theory of electromagnetism (*see* ELECTRICITY & MAGNETISM p.20). "Maxwell's equations", as normally taught, are in fact Heaviside's creation — Maxwell's corresponding mathematical expressions are more awkward (he did not use vector notation), and scattered over different sections of his celebrated *Treatise*.

Heaviside came from a poor family and grew up in dismal surroundings in Camden Town in London, without much formal education. He continued poor all his life, prizing independence above all else, spurning most offers of employment and even a Royal Society grant. He lived in Paignton (and later in Torquay), the home of his parents, because of its seclusion, far from centres of intense scientific activity. One of the people who helped him was his uncle, Charles Wheatstone, the inventor of the working (cable-transmitted) telegraph, who gave him an early job as telegrapher when he desperately needed an income. Heaviside continued to focus much of his work on signal transmission ever after and is well known to radio enthusiasts for his assertion (substantially correct) that there must be an "ionosphere" — an electrically conducting layer in the outer atmosphere — to account for the long range of some wireless signals.

Paignton has no explicit memorial for Heaviside, but a historical guide (published 1974) lists him among the town's premier citizens. It identifies his residence from 1889 to 1897 as Palace Avenue, where traces of a former peaceful residential area can in fact still be seen: solid Victorian houses, a park with flower beds and benches to sit on, and an old theatre at the end of the street.

PENZANCE (Cornwall)

Penzance, virtually the end of the world for England (the literal Land's End is just a few miles away) is a delightfully situated seaport, at the centre of what was once a thriving mining area, the world's premier supplier of tin for almost two thousand years and (more briefly, in the nineteenth century) a source of copper. Humphry Davy, one of England's greatest chemists (*see* CHEMISTRY p.44), was born here in 1778. Penzance celebrates the fact with a statue at the top of Market Jew Street, which rises steeply from the harbour to the Market House. The statue is elevated on a pedestal, augmenting the dominance of its position, and bears a plaque with a scientifically accurate summary of Davy's career.

Cape Cornwall. It is appropriate to advise the tourist to avoid Land's End, the unique position of which has been shamefully exploited for commercial profit. On the other hand, we recommend a visit to Cape Cornwall, only five miles away, with a good view of Land's End and situated in an area rich in abandoned tin mines, whose brick smoke-stacks stick out into the air all around us. It should be remembered that Humphry Davy, aside from his purely scientific accomplishments, was the inventor of the miner's safety lamp, a device not needed in a tin mine, but possibly inspired by his origins in a mining area.

The Levant steam engine and related mining exhibits are a National Trust facility at Pendeen, just north of Cape Cornwall on the B3306. It is open July to end of September, Sun–Fri, 11–4, with more limited opening from Easter to June. Phone (01209)-216657.

PLYMOUTH (Devon)

Plymouth boasts two of the proudest achievements of British civil engineering, Smeaton's Eddystone light tower and Brunel's Royal Albert bridge across the Tamar river into Cornwall.

Plymouth Hoe, high up above the Sound, is a popular place for tourists, the place where in 1588 Francis Drake (at least according to legend) insisted on finishing his game of bowls before setting out to tackle the Spanish Armada. In good weather one can see the Eddystone Light from here, 14 miles out at sea, one of the oldest and most famous of all lighthouses. The first lighthouse was built here in 1698; the present version dates from 1882; the most celebrated is John Smeaton's, which survived the seas for 120 years, from 1762 to 1882, and was replaced only because the rock on which it stood had begun to crack. (*See* LEEDS, Yorkshire, where Smeaton lived, p.230.) The tower was re-erected on the Hoe in 1884 and stands there today as a Smeaton monument.

The Royal Albert bridge (completed 1859), like George Stephenson's Menai bridge in Wales before it, used hollow iron tubes as weight-bearing elements, with enormous benefit for strength/weight efficiency — there is a difference in that Brunel's tubes were oval in cross-section, whereas Stephenson's were rectangular. The bridge is nearly half a mile long and the Admiralty required a 100-foot headway underneath; it has carried rail traffic without interruption to this day. The bridge is seen to good advantage from the A38, crossing the river out of Plymouth on its own adjacent suspension bridge. It bears the inscription "I.K. Brunel, Engineer, 1859", in bold letters high above the supporting arches.

SIDMOUTH (Devon)

At the Norman Lockyer Observatory in Sidmouth we celebrate a man of independence and spirit and a foremost figure in astrophysics. His principle achievement was that he discovered the existence of the chemical element helium;

what is remarkable is that the discovery used no chemical methods at all, but was based on an anomalous dark line in the solar spectrum, arising from the solar corona and visible only during a solar eclipse. Lockyer understood that this line must arise from atomic absorption of sunlight of a single frequency — the frequency per se could not be assigned to any known atom and thus indicated the existence of the new element, which he named "helium" from the Greek word for the sun. Helium was ultimately discovered on earth (as a product of radioactive ores) by William Ramsay — over 25 years later — and prepared in pure form by him and his collaborators.

Norman Lockyer (1836–1920) was born in Rugby, where his father was a keen amateur scientist, cofounder of the local literary and scientific society. Norman, too, was initially an amateur, lacking formal training in science, but he became an expert in solar astronomy, especially on the then anomalous features, such as sunspots and the corona. The recognition of solar helium (1868) required a deep comprehension of physics (spectroscopy) and chemistry as well as astronomical observation. Is there a lesson for us today in the fact that Lockyer possessed the appropriate learning without explicit "training" for the job? Lockyer lived in Sidmouth after he retired and privately financed his observatory here in 1912 (aged 76) after the government had refused to provide one for him. His work late in life included a book about Stonehenge's possible function as an astronomical observation post. He calculated that the sun would have been in exactly the right position to create the celebrated alignment of the rising sun with the circle's axis on midsummer day of 1680 B.C. and therefore suggested that as the date of erection, with a margin of possible error of a couple of hundred years.

Lockyer's observatory, situated on Salcombe Hill above Sidmouth, has been rescued from decay and collapse by public-spirited citizens and the East Devon District Council. There are three separate domes; two of them retain their original telescopes and the third has a more modern instrument. They are used by the local schools for instruction and they are open to the public on designated nights. The local District Astronomical Society holds its meetings here and sponsors public lectures; the Amateur Radio Society meets here, too — scientific activity (in contrast to the more common passive role played by museums) is at an exceptional level.

Open "skywatch" sessions are held at the observatory 2 nights a week. The grounds are always open: there are 2 car parks, a picnic area, and 'splendid views of the Devon coast. Phone (tourist office): (01395)-516441.

STALBRIDGE (near Sherborne, Dorset)

Robert Boyle lived here as a youthful master of the manor (Stalbridge Park) from 1645 to 1655 — the house had been built by his father, the Earl of Cork, in 1638. Boyle had recently returned from his grand tour of Europe and was temporarily in somewhat dire financial straits because the Irish rebellion had cut off (for the entire Boyle family) the accustomed flow of rent income from their Irish estates. It is tempting to think that this was where Boyle found his niche in life and that the seeds of his rational natural philosophy were planted here, but historians of science can find no evidence for that. He wrestled with personal problems and ethical questions, and found his vocation in the sense that he became determined to lead a *useful* life, but there is no evidence for an interest in science until after he left Stalbridge for Oxford. He was over 30 years old by then, late in life by modern standards to embark upon a career in the physical sciences.

Stalbridge Park is at the northern edge of the village, on the A357, next to a church. Little remains of the mansion; even the lion-crested stone pillars at the entrance were probably erected after Boyle's residence. The perimeter wall, however, is partly original and testifies to the considerable size of the property.

STONEHENGE (near Amesbury, Wiltshire)

Here we have Britain's most famous prehistoric monument, an enigma to some, a place of mythical significance to others, and (for the inquisitive) an object for speculation and research. In the realm of science there are three major questions.

(1) Who were the people who had the skill and determination to build this complex structure? John Aubrey (around 1650) casually suggested the Druids, a Celtic priesthood who lived here in Roman times, but that is certainly wrong, for Stonehenge is much older than that — it was built in several stages, beginning about 3000 B.C.

(2) What was Stonehenge's function? The celebrated alignment with the rising sun on midsummer day has led to speculation about quite sophisticated astronomical knowledge and use of the stones for charting the movement of heavenly bodies. In fact, the alignment is probably fortuitous and comparison with the design of other stone circles in Britain does not lend much support to the speculation. Which does not deny the likelihood that the structure, once

England's most famous ancient monument. The smaller rocks in front are the bluestones that came from the Preseli mountains in Wales.

140

built (for whatever principal ceremonial purpose), did indeed serve as a crude calendar, just as we might today use a tree in our garden to follow the progress of the sun's position through the seasons.

(3) Where did the stones come from? The outer circle, huge sarsen stones with horizontal lintels, has a local origin, but the next circle within consists of bluestones, which without question came from the Preseli Mountains in southwest Wales. Why did the Stonehengers go so far to get their rocks? What route did they take? How were they carried? Current research (*see* PRESELI, in Wales, p.266) suggests that men did not carry the stones at all, but that they were carried by the flowing glaciers of an early ice age and were lying about here, ready to use, when stone age people arrived on the scene.

Some of the myths can clearly be discounted; much of the mystery remains. Visitors who want to draw their own conclusions are advised not to rely on Stonehenge alone, but to look also at the stone age circles in Avebury and (especially) those farthest to the north, on the Scottish offshore islands.

Stonehenge stands right beside the A303, one of the major highways to the southwest. It is open to visitors all year, 9.30–6.30 (summer), 9.30–4 (winter). Phone Bristol (0117)-973-4472.

TRURO (Cornwall)

Truro, the seat of Cornwall county council, is a latecomer to the status of a cathedral city: the building was not completed till 1910. It is conspicuously unlike any other British cathedral in that its stained glass windows pay tribute to philosophers as well as saints, including for example Erasmus, Francis Bacon and even Isaac Newton. It is worth spending a few minutes to read the tributes to these early scholars (on posters found next to each window), with their unambiguous assertion of a lack of conflict between religion and the sciences.

Not surprisingly, the cathedral contains a tribute to Cornwall's famous astronomer, John Couch Adams (*see* LAUNCESTON p.130), in the form of a commemorative tablet in the north transept. The inscription is in Latin, but there is a translation below: "Tracing his way by the sure clue of mathematics," it tells us, "through the boundless night of space, he found the uttermost of the planets" — and then it goes on to tell us about his modesty and love of God.

WIMBORNE MINSTER (Dorset)

The town is named after the church. One of the noteworthy sights in the latter is an astronomical clock, a time keeper and simultaneously a model of the motions of the earth, sun and moon. The exact date of construction has not been established — it might have been as early as 1320 — but it was certainly in place by 1409. This was long before Copernicus, whose heliocentric system dates from 1543, and the clock thus follows the Ptolemaic system, with both sun and moon revolving about the earth. The sun is in effect the hour hand of the clock and there is no separate minute hand. The moon revolves against a starry sky on a smaller orbit, but also turns about its own axis to mark the moon's phases. The clock has been redecorated, as faithful to the original as can be ascertained. The mechanical works are newer, dating from 1792.

The lesson in all this is that the Ptolemaic system can accurately reproduce these motions as we see them from the earth and could equally well account for the motions of the planets as well — and any number of other models could do the same. Copernicus knew this, of course. He never proved or even asserted that the sun is the centre, but simply demonstrated that far fewer assumptions about individual motions were needed for a heliocentric system, fewer equations in mathematical terms, fewer gears in a clock such as we see here. An early application of Occam's razor (*see* OCKHAM, Surrey p.112)?

The astronomical clock is inside the church in the West Tower. The same machinery drives a *quarter jack* on the external north wall (installed 1612), where a figure strikes two bells every quarter hour. The figure used to be a monk, but was replaced by a soldier after the Napoleonic war.

The minster is open daily for sightseers, 9.30–5.30, except during services.

4
Midlands (South)

ALDBURY (near Berkhamsted, Hertfordshire)

Here, in an unlikely forest setting, we have a monument to a member of the Bridgewater family, a family that can lay claim to one of the giants of the industrial revolution, but may have had equal influence on science through a later, less conventional scion.

The monument is on the Ashridge estate, 4000 acres of unspoilt open spaces, commons and woodlands, which used to belong to the family, but is now a National Trust property, a fine place for walking and wildlife observation. The monument is an imposing column with a viewing platform at the top, erected in 1832 to honour Francis Egerton, 3rd Duke of Bridgewater (1736–1803), acclaimed in the inscription as "Father of Inland Navigation". He was responsible for the building of grand ship canals in the Manchester area, which we have described in our entry for Manchester (see NORTH OF ENGLAND p.236).

But we should also remember the Duke's younger brother, the 8th Earl of Bridgewater (1756–1829), who was a naturalist, antiquarian and defender of traditional religious beliefs. In the latter role he made a bequest of £8000 for the writing and publication of a treatise "On the Power, Wisdom and Goodness of God, as manifested in the Creation", but the curious thing is that he made the bequest not to a religious

The Bridgewater Monument on the Ashridge estate near Aldbury.

foundation, but to the Royal Society, and he gave them the power to select suitable authors. Eight authors were duly chosen, for their scholarly and/or scientific approach to the subject, and they were given an honorarium of £1000 apiece. Some of the volumes have lasting interest and importance, e.g., *Astronomy and General Physics considered with reference to Natural Theology* by William Whewell. Perhaps the most intriguing title is *Chemistry, Meteorology, and the Function of Digestion, considered with reference to Natural Theology* by William Prout. Prout (*see* HORTON, Avon p.127) is well known to chemists and physicists for his far-sighted proposal (Prout's hypothesis) that all atomic weights are integral multiples of the weight of a hydrogen atom, but, as the title of his volume indicates, that was not his only scientific interest.

The Bridgewater Monument and National Trust Information Centre lie off the B4506, about 4 miles north of Berkhamsted. Ascent of the tower (for an overview of the estate) is possible April to October, daily except Friday, afternoons only.

BERKELEY (Gloucestershire)

Berkeley is the home of Edward Jenner, who observed that cowgirls never contracted smallpox and worked out the method of vaccination against the disease to protect the rest of us. The town is about halfway between Gloucester and Bristol, far enough away from the main road to have remained a quiet little market town. Teashops outnumber service stations. Little old ladies wander around the town, visiting the historic castle and the ancient church of St. Mary the Virgin. Edward's father (Stephen) was vicar of the church from 1729 to 1755 and Edward was born in the vicarage — later on, when he could afford it, he purchased the Chantry, a fine mansion adjacent to the church, which remained his home for the rest of his life. It now houses the Jenner Museum (established in 1985).

The museum contains all kinds of memorabilia and includes a picture of the humble "old vicarage" where Jenner was born. The exhibits show papers dating from the days when Jenner was a naturalist, including one on the unfriendly nesting habits of the cuckoo and another on bird migration. Jenner's old study is preserved and contains some nineteenth century surgical instruments, including one of the multi-pointed devices (for penetration below skin surface) which were used for vaccinations from his day until well into modern times. There is a separate "all about smallpox" room, with a gruesome picture of someone ill with smallpox, covered densely from head to toe with huge ugly pustules. All of Jenner's medals and honours are here and copies of contemporary humorous cartoons.

Another room is dedicated to the World Health Organization. Short films are shown here on the nature and course of smallpox and on its worldwide eradication by the

WHO-sponsored vaccination programme. One film is done in cartoon form for children, detailing Jenner's life and work. On the outside are pretty gardens and tucked away in a corner is a little thatched hut, the *Temple of Vaccinia*. It was built for Jenner by an eccentric friend and poor people came there to be vaccinated free of charge. Inside are the shoulder blade and scapula of a whale, reputedly dissected by Jenner in 1788.

Next to the Chantry is the Church of St. Mary the Virgin, parts of which date back to the twelfth century. The chancel of the church, at the east end, is where Jenner and members of his family are buried. The stained glass east window of the church was put in explicitly in honor of Edward Jenner in 1873. It shows scenes from the Bible of Jesus healing the sick, mostly by the laying on of hands — there is nothing that has to do specifically with smallpox or Jenner. (It is quite different in that respect from the Pasteur mausoleum in Paris, where the decorations are exclusively in praise of Louis Pasteur!) Other local celebrities are buried inside the Berkeley church, including Thomas, Lord of Berkeley (1326–1361), who was the lord of the castle at the time when Edward II was murdered there in 1327.

The Temple of Vaccinia, where Jenner gave free vaccinations for those who could not afford to pay.

The museum is open from April to September, Tue–Sat, 12.30–5.30, Sundays and Bank Holidays, 1–5.30; also Sundays in October. Phone (01453)-810631.

CHURCHILL (near Chipping Norton, Oxfordshire)

William Smith, the father of British descriptive geology, was born in this village in 1769, the son of a blacksmith. His future career was at first determined quite accidentally, when he was hired as a helping hand by a surveyor from Stow-on-the-Wold, who was making a survey of the parish in preparation for the enclosure of common lands. Smith proved to be an intelligent lad and was taken into the business. Five years later he was sent out by his employer on his own, to survey an estate near Bath, and it was there that he began to think about the succession of strata and to begin his slow conversion from simple workman to acclaimed professional (*see* HIGH LITTLETON, Avon p.126). Churchill marks Smith's origins here with a rough stone monument on the village green.

Churchill is 3 miles southwest of Chipping Norton, on the B4550. Warren Hastings, famous empire builder, is another village native (born in 1732).

ECTON (near Northampton, Northamptonshire)

The village of Ecton was for centuries the home of the family of Benjamin Franklin (1706–1790), American patriot, whom we don't usually think of as English at all, but he was 70 years old by the time he signed the Declaration of Independence, and before then had never questioned his English nationality. When, like many another English gentleman, he became an amateur scientist in Philadelphia and made important discoveries about electricity, he turned to the Royal Society to have them published, was soon nominated to become a fellow and even won the coveted Copley Medal in 1754. Later on he was sent to London by the colony of Pennsylvania as a petitioner (not in any sense as a threatening separatist) and he lived there from 1757 to 1762 and again from 1765 to 1775. He was a fully accepted member of the scientific community, an active participant in Royal Society meetings and member of many of its committees; he continued to do experiments of his own, such as the well-known wave-stilling experiment on Clapham Common. Franklin visited Ecton in 1758 to search for his family roots, but the last of the local line, his uncle Thomas, had died in 1702 and no close relatives remained.

The Franklins had been the Ecton blacksmiths, the business passing from father to oldest son. The smithy was at the back of what is now the Three Horseshoes Inn — it burned down long ago, but parts of the old structure remain in the inn's backyard. The old family home, on the opposite side of the street, was sold after Thomas died and used as a school — there is an inscription above the door. Thomas and his wife Eleanor are buried in the yard of St. Mary Magdalene church (near the north door) and there is a tablet to indicate the connection to their famous American nephew.

Ecton is 5 miles east of Northampton off the A4500.

EDGEHILL (near Stratford-on-Avon, Warwickshire)

William Harvey was King Charles I's physician and friend. At the first major battle of the Civil War, the battle of Edgehill in 1642, the king's own sons (the future Charles II and James II) were still in the royal entourage and were put in Harvey's charge as they watched the battle from the sidelines. Today we get the best view of the battlefield from the Castle Inn on top of the escarpment and this is also where pamphlets and other information about the battle can be obtained. One would think (taking the modern view) that this is also the place where the spectators stood, a safe distance from the bullets, but history tells us otherwise — the princes would not have tolerated such cowardice. They were probably behind some clumps, on the road at the foot of the escarpment, near the village of Radway. (The battle itself was inconclusive and the ultimate doom of the king's cause not yet apparent.)

Edgehill is a mile north of the A422, about halfway between Stratford and Banbury. The Castle Inn is signposted.

GLOUCESTER (Gloucestershire)

The city of Gloucester is the birthplace of Charles Wheatstone (1802–1875), a versatile physicist, somewhat of a dilettante, but always full of ideas for new experiments and inventions. His father was a music seller, his uncle was a musical instrument maker, and Wheatstone's work in physics began (appropriately) in the field of acoustics. But he also experimented with electricity, where his most fruitful work was to measure the speed of electrical conductance in a wire, getting a result that was not very accurate, but sufficient to show that conductance is very fast. This gave him the brilliant idea to use electric current for transmission of messages, resulting in the first recorded patent for a telegraph, awarded to him in 1837. Wheatstone also invented the rheostat and his name is known to every physics student for the "Wheatstone Bridge" circuit for measuring electrical resistance. (History shows that he was not the actual inventor of the latter, but he certainly popularised it.) Professionally, despite the fact that he lacked formal scientific education, he was appointed professor of experimental physics at King's College, in London, a position he held from 1834 to 1875. He was elected a fellow of the Royal Society, and received many other honours.

Gloucester's City Museum has a "Wheatstone Hall" at the back, but nothing other than the name to connect it with the man. The city intended a proud memorial in the form of a statue — a 20-inch model was built, showing Wheatstone in academic gown, his hand resting on a dial telegraph, but the funds to build it were never forthcoming. (The model survives and can be seen is in the museum keeper's office.) A brass memorial plaque exists at the Gloucestershire College of Arts and Technology, in a not very prominent place behind the reception desk.

Gloucester's major tourist attraction is its great cathedral, with its gorgeous tombs of Edward II and other noble personages. It also contains, close to the public entrance, a large and fanciful statue of William Jenner, the discoverer of immunisation against disease, who hailed from nearby Berkeley (*see* p.145).

Several biographies state that Wheatstone was born in a fancy Gloucester suburb, but it is now thought that the actual birthplace was above the family shop, a building at 52 Westgate Street in the city's pedestrian precincts. The City Museum and the Technology College are on Brunswick Road, just off Eastgate (also pedestrian precinct).

HARPENDEN (Hertfordshire)

Agricultural research at the Rothamsted Experimental Station is the attraction here. The institution was conceived by John Lawes (1814–1900) in 1843 — it is the place where he and his associate Henry Gilbert established the true foundations of the science on which humanity depends for its food supplies.

The German chemist Justus von Liebig probably provided part of the motivation. He was the doyen of organic chemistry, the ultimate authority, but he was wrong when he tried to apply his knowledge to agriculture. Liebig asserted that plants could get all the nitrogen they need from the air, which Lawes and Gilbert demonstrated to be false. To supply other elements, Liebig patented and sold a mineral fertiliser with essentially *insoluble* constituents, the intended purpose being to prevent them from wastefully flowing away, the actual result being to make them unavailable to plant roots. Lawes and Gilbert demonstrated that fertilisers need to be *soluble* to be available for use — it is difficult to imagine today that one could ever have thought otherwise — and reinforced this with the finding that dissolved minerals are absorbed by soil particles and don't "float away". All subsequent advances in crop fertilisation depend on these principles.

John Lawes set up the Rothamsted Experimental Station on the family estate he had inherited when his father died. He instituted the so-called "classical" experiments here, where the effects of nitrates, phosphates and other nutrients were quantitatively measured — some of the same experiments continue to this day, on the same strips of land, but they are now supplemented by vastly more sophisticated methodology and instrumentation. Greenhouses and chemical laboratories abound; genetics and environmental concerns have become as important as crop fertilisation; a separate department (established more than 70 years ago by the renowned statistician R.A. Fisher) handles computer programs for the analysis of results. Given the benefits that all of us derive from the work done here, one would think that Rothamsted should receive more publicity than it does.

Rothamsted is a working institution, with no facilities for casual visitors, but public footpaths cross the estate and a glimpse of fields and of what goes on in the greenhouses can be seen from them. A 15-minute walk takes one to the original manor house and farm where Lawes was born, which is now a private residence for students. A stone monument was set up at the entrance to the station in 1893 to commemorate 50 years of experiments here, "the first of their kind in agriculture", and a supplementary plaque was added in 1993 to celebrate another 100 years. Just inside the gates is a large map of the area, showing the location of footpaths.

The entrance to the Experimental Station is at the edge of Harpenden, just off the main road to St. Albans, adjacent to the pub "Silver Cup".

MINCHINHAMPTON (near Stroud, Gloucestershire)

James Bradley (1693–1762), the Astronomer Royal who first established the Greenwich meridian, was a Cotswold man. He was born in the village of Sherborne, not far from Stow-on-the-Wold and other popular Cotswold tourist sites. He married (rather late in life, in 1744) a lady from Chalford (near Stroud) and retired there when he felt he was no longer fit for work. He died in Chalford and was buried in the parish churchyard in nearby Minchinhampton.

A brass plate in the Lady Chapel of the church (south transept) commemorates the astronomer. It was originally on the tombstone in the churchyard, but 19th century vandals tried to steal it several times and it was therefore deemed prudent to transfer it within the church. The stone at the actual grave now bears a simple inscription cut into the stone.

Minchinhampton, 5 miles southeast of Stroud, is best noted for its market place. The church is normally kept open during the day — 20th century vandals in this case appear to be less of a danger than their 19th century counterparts.

NORTHAMPTON (Northamptonshire)

Doddridge's Academy in Northampton was one of the so-called "dissenting academies", which began to be formed in the late 17th century and came into the open after the accession of William and Mary in 1689. They trained men for business and the professions with a far wider and more up-to-date curriculum than that of grammar schools and universities: they included mathematics and science, and, when scientific invention revived in the 18th century, the impetus came from these academies and from individual craftsmen, not from the universities or the Royal Society. Doddridge's academy is historically the most important in relation to the subjects of science. Here is where methods and principles were developed that were put to use by daughter academies in Daventry and Warrington — both explicitly related to science through their association with Joseph Priestley. (*See* WARRINGTON, Cheshire, p.188.)

The educational mission of the academies was closely linked to the practice of non-conformist religion. Philip Doddridge was preacher here from 1729 until his death in 1751, at what is now called the "Doddridge and Commercial Street United Reformed Church", but used to be "Castle Hill Meeting" when first opened in 1695. There is a fancy memorial to Doddridge inside, above the altar, a plaque on an outside wall of the church, and a portrait in the minister's study. The academy itself quickly needed large quarters to accommodate a growing student body, and they were provided by the Earl of Halifax (Lord Lieutenant of Northamptonshire), who donated his town house for the purpose. The house is still there, at the corner of Sheep Street and Greyfriars, marked by a plaque.

The city of Northampton takes impressive pride in this

aspect of its history. Note especially the modern murals in Grosvenor Centre (the heart of "downtown"), one of which features Doddridge and a group of choristers. Other places of interest in connection with Doddridge are part of a "Doddridge Trail", described in a pamphlet available at the church or the town tourist office.

The historic murals in Grosvenor Centre are on the upper floor, in the North Piazza, opposite the entrance to Sainsbury's. Doddridge's Church is on the north side of Mare Fair, close to the railway station. On the opposite side of Mare Fair is the establishment church, the Parish church of St. Peter, which dates back to the year 1160 — the great geologist William Smith (*see* CHURCHILL), who died in Northampton in 1839, is buried within, p.147.

Daventry (Northamptonshire). Doddridge named Caleb Ashworth to be his successor. The latter had been minister at the dissenting church in Daventry, 13 miles from Northampton, and this meant that the Academy moved there, too: it was named "Doddridge Academy" in honour of the deceased founder and is historically important in its own right as the place where Joseph Priestley was a pupil. The building still stands (on Sheaf Street) and is marked by a stone plaque on the front.

OXFORD (Oxfordshire)

Oxford is the cradle of the Royal Society, the oldest scholarly society in England and the most influential of all of Europe's science academies — it started informally at Oxford as a discussion group sometimes dubbed the "invisible college", but then moved to London where it was formally constituted with the blessing of Charles II. Robert Boyle (1627–1691) and Robert Hooke (1635–1703) were among the most prominent of the early Fellows of the Society and did some of the work for which they are famous here in Oxford. The great astronomer Edmund Halley (1656–1742) was Savilian professor of geometry and did almost all his scientific work in Oxford — he was 60 years old by the time he was appointed Astronomer Royal and moved to the Greenwich Observatory. The heady atmosphere of those early days has not been matched since then and Oxford is generally ranked below Cambridge in the more modern history of British science.

Specific entries for Oxford are grouped as follows:

1. Oxford Colleges
2. University Science Departments and Laboratories
3. Museums
4. Other Places of Interest

---

1. Oxford Colleges

Christ Church - Alice in Wonderland.   Christ Church is Oxford's largest and grandest college, but relatively new by Oxford standards, having been founded in the reign of Henry VIII. One of its former students, Thomas Willis (1621–1675), was an early member of the enthusiastic scientific group (Boyle, etc.) mentioned earlier, and did experiments at the college while still a student. He drifted away from the others because his chief interest was in the practice of medicine, but he continued original work as well — he published the then definitive work on the anatomy of the nervous system in 1664 and is best remembered for his classification and functional definition of the human system of cranial nerves, which remained standard doctrine for over 100 years. The college takes no particular notice of his one-time presence: he is included in one of the stained glass windows in the Great Hall, along with some other medical men.

Somewhat more recognition is given to Lewis Carroll (pseudonym of Charles Dodgson, 1832–1898), whom we all know as author of *Alice in Wonderland*, but who was also Christ Church's tutor of mathematics and logic. The working of his analytical mind is reflected everywhere in his writings. One of our favourite passages is from "The Hunting of the Snark":

> Just the place for a Snark! I have said it twice;
>   That alone should encourage the crew.
> Just the place for a Snark! I have said it thrice;
>   What I tell you three times is true.

which is of course (presumably intentionally) the exact antithesis of the logic of science, where no amount of forcible authority has any weight compared to the evidence of experiment. Lewis Carroll had personality problems which made him unable to befriend or relate socially to adults. The _Alice_ books were based on stories with which he entertained the three young daughters of Christ Church's Dean Liddell, whom he used to take boating on the river Thames, setting out from below Folly Bridge — just a few steps from the College.

One of the stained glass windows in the Great Hall is devoted wholly to Lewis Carroll, illustrated with representations of characters from his books and one of Alice Liddell herself. There is also a posthumous portrait of the author near the Hall entrance: "Student, 1852–1898", is how he is described, which is not indicative of failure to pass exams, but just an old College custom, that we are considered students for ever, no matter what our rank.

Christ Church College and Cathedral are a single linked institution. It caters more to tourists than other Oxford colleges, open 7 days a week, with an admission charge and official guides. The Cathedral and the Great Hall are part of the tour, except at meal times, when the Hall is in use as the student refectory.

Linacre College. Linacre College is one of the newest, first established in 1962. It is also thoroughly modern in outlook: it admits only graduate students and it promotes socially significant lectures. The college is named, however, after one of Oxford's most ancient sages, Thomas Linacre (1460–1524), who went to Italy right after his student years (probably for about 6 years in the 1480s) to sop up Italian renaissance culture and became a figure of prime importance on his return, a teacher of other scholars (Thomas More, Erasmus) and at the courts of Henry VII and Henry VIII. He was also a physician and the founder of the Royal College of Physicians, which initially had its quarters in Linacre's own home in London (_see_ ROYAL COLLEGE OF PHYSICIANS, London, p.58). The College possesses some of Linacre's books and there is a line drawing of him in the reception room.

Most college locations can be found on every Oxford tourist map, but Linacre is usually missing. It is at the corner of St. Cross

Road and South Parks Road, next to the entrance to the Parks.

Merton College.   Merton College, one of Oxford's oldest, was founded in 1264. In the antechapel, on a wall, is a fancy monument to Sir Henry Savile, who founded the Savilian professorships of astronomy and geometry that have been held by so many distinguished Oxonians of the past. Henry Briggs (Briggsian logarithms, *see* WORKING WITH NUMBERS p.15), who held the geometry chair from 1619 to 1630, has a contrastingly modest memorial. He is buried below the floor of the antechapel and the stone bears only his name ("Henricus Briggius") — not even the dates are given. A familiar name on the list of Wardens of Merton (to be found on an antechapel wall) is that of William Harvey, but its presence here is misleading. King Charles I had made Oxford his headquarters for fighting the war against the rebel republicans and he ordered Merton to dismiss the incumbent and appoint Harvey in his place. They didn't like the idea, but had little choice — Harvey's predecessor was restored after less than a year, as soon as the war was lost.

   Merton's library is the oldest in the country and contains (in a glass case) an astrolabe believed to have belonged to Geoffrey Chaucer, the poet who wrote the *Canterbury Tales*. Chaucer (1340 to 1400) wrote a treatise on the astrolabe and his tales provide frequent evidence of his familiarity with matters scientific. He was never a member of Merton and nobody knows why the astrolabe resides here.

Merton College has set hours for visitors: Mon–Fri, 2–4, plus Sat at the same times from March to October. Access to the library is limited — inquire at the porter's lodge.

Somerville College.   Somerville is one of two women's colleges founded in 1878/79. It was named after Mary Somerville (1780–1872) an extraordinarily independent lady from Jedburgh in Scotland, who wrote respected books on physical science and astronomy — including a translation of Laplace's *Celestial Mechanics* from the French. She managed to retain her feminine charms as well and was once known as the "Rose of Jedburgh". Dorothy Crowfoot, 1964 winner of a Nobel Prize for determining the structure of insulin, was a student here. So was Margaret Thatcher, who was briefly a chemist before entering the more uncertain world of politics.

Wadham College.   Wadham College is where the group that proved to be the embryo of the Royal Society was founded in 1648. John Wilkins, the Warden of the college (later to become Bishop of Chester), was the guiding spirit and typifies the early days of the Royal Society, which reflect a unique _voluntary_ effort, an inner-directed enthusiasm for science and its prospects by rank amateurs as well as by persons whom we would today consider "professionals" in the sense that science was their full-time occupation. Wilkins was the author of a curious book, _The Discovery of a World in the Moon_, with an appendix on "The possibility of a passage thither". Pleasure and profit will be as great, he claimed, as the pleasure and profit derived from the discovery of America. (The Duchess of Newcastle asked where to rest her horses if she undertook the journey. "Use your castles in the air," Wilkins is said to have replied.) Also among the twelve founding members of the Society was Christopher Wren, architect of St. Paul's and other fine churches, who was a student at Wadham.

Strangely, Wadham College does not choose to rejoice in its early incisive role. The Royal Society has created a research professorship at Oxford in commemoration of its birthplace and the appointee is by statute a fellow of Wadham — that's the only memorial there is.

2. University Science Departments and Laboratories

The centres for today's science teaching and research are mainly clustered along Parks Road and South Parks Road and most of them have little in them to interest visiting lay people. An exception is the Physiological Laboratory. Its most distinguished professor was Charles Sherrington (1857–1952), professor at Liverpool from 1895 to 1912 and here at Oxford from 1913 to 1935, who was a pioneer in research into reflex physiology and the relation between anatomy and function of nerves. He introduced a substantial fraction of the now basic vocabulary of neuroscience —

such as synapse, motoneuron, excitatory and inhibitory states. He was a Nobel laureate for physiology and medicine in 1932, while at Oxford.

The department has a small lecture room named after Sherrington, and (in the hall outside it, opposite the entrance to the library) a glass case displaying some of the equipment he used. The walls of the staircases and hallways are lined with pictures of famous physiologists of the past, with accompanying brief biographical summaries — they go back to Thomas Linacre, Thomas Willis and William Harvey and continue to the present century. Most of the names may not mean much to lay people, but the pictures will provide a rich source of enjoyment for anyone with even a smattering of familiarity with the subject. (A painting at the foot of the stairs, showing Sherrington with his most famous pupil, John Carew Eccles from Australia, is particularly eye-catching. Eccles went on to share in a Nobel Prize in 1963 for his own work in neurophysiology.)

The Earth Sciences Department merits brief mention for the fact that it displays in its library a medal presented to a relatively obscure geologist, Joseph Prestwich (1812–1896), and worn by him at a convention in London in 1888. Prestwich, Oxford professor from 1874, contributed significantly to our appreciation of the great antiquity of the human species. The crux of the matter was a long-standing claim by a rank amateur in France, Jacques Boucher de Perthes, to have discovered man-made tools in the sands of the estuary of the Somme, near Abbéville, in debris that also contained elephant and rhinoceros bones that could have been as much as a million years old. Boucher had been ignored and even ridiculed, partly because he never found actual human remains in the area. Prestwich thought the matter was too important to dismiss out of hand and went to Abbéville himself. He confirmed Boucher's findings and himself discovered human bones. Added to an equally suggestive find in Brixham (Devon), this established the antiquity of the human species — a million years or more. It has never been questioned since.

Oxford's most famous (at times even infamous) geologist was William Buckland (1784–1856), who is mentioned in several places in this book. His portrait hangs in the Earth Sciences Library.

3. Museums

Museum of the History of Science (Old Ashmolean Building). Here we are steeped in the history of the University just as much as the history of science. The building was erected between 1679 and 1683 to house lecture rooms and a chemical laboratory as well as an upstairs museum. The old entrance signs are still there — *SCHOLA NATURALIS HISTORIAE* on the ground floor and *OFFICINA CHIMICA* at the basement end of the stairs. Buckland lectured here, in the very room we now see — a lithograph shows him doing so in 1823, to an audience consisting mostly of mature dons and including some well-known figures, such as the Conybeares. Thomas Beddoes (*see* BRISTOL p.120) was briefly a reader in chemistry at Oxford and worked in the basement of the building. A glass cabinet shows some of his equipment, and nearby is a huge burning lens used by him — it was a common way to heat things at the time.

Entrance to the Old Ashmolean Building.

Other exhibits relate to work done at Oxford, but not in this building. One relates to Henry Moseley (1887–1915), the brilliant chemical physicist who was killed (aged 27) at the battle of Gallipoli. Moseley was a student at Trinity College, then worked for a while with Rutherford in Manchester and William Bragg in Leeds. He deliberately chose to become independent (convinced that Rutherford had no more to teach him) and came to the Clarendon physics laboratory at Oxford. Here he discovered the single most important link between chemistry and the then rapidly gowing science of subatomic physics, a precise relationship between the frequencies of characteristic x-ray emissions of the elements (induced by bombardment with electrons) and the order in which elements appeared in the periodic table. It unequivocally established the *atomic number* as the linch-pin of chemical behaviour, demolishing at a stroke all controversy about inconsistencies between atomic weight and position in the periodic table. The apparatus used by Moseley is on display in the museum.

An exhibit on penicillin (*see* ALEXANDER FLEMING MUSEUM, Paddington, London p.85) is also of interest, relating to the work done at Oxford (in the 1940s) to purify penicillin and to determine its chemical structure. Bedpans from the hospital were used at first to grow the *Penicillium* mould. There is a display featuring Dorothy Crowfoot, the former Somerville student, who led the team that determined the 3-dimensional details of penicillin's molecular structure by use of x-ray crystallography.

In its more general collection the museum has the world's best collection of astrolabes and other ancient astronomical instruments, and there are early mathematical devices as well, including an example of the circular slide rule designed by Oughtred in the 17th century. From the same era, there is a replica of Boyle's air pump, with the aid of which Boyle's law for the relation between gas pressure and volume was determined.

The museum is open Mon–Fri, 10.30–1 and 2.30–4.

University Museum. The University Museum (on Parks Road) is a Victorian architectural extravaganza, with a huge central open space which dwarfs even the dinosaur skeletons that are scattered about it. It has a glass roof, supported by arrays of ornate cast iron pillars. Around it are two arcades, one above the other, the upper forming a gallery around the central court, with cloister-like arches and columns. The columns are made of highly polished stone, each one different, representing in effect a shining collection of British rock samples, all meticulously identified — there are more than a hundred, assembled here as a labour of love by John Phillips (_see_ YORK p.254), geologist and first keeper of the museum.

When first opened in 1860, the museum housed all the university departments of science and even contained a chemistry laboratory, but the present museum is wholly for display, dedicated to geology, zoology, entomology, and other aspects of natural history. The exhibits themselves are supplemented by a multitude of statues (and some busts), seemingly chosen without much discrimination as to where and when their subjects lived or exactly what they did: Aristotle, Harvey, Roger Bacon, Francis Bacon, Newton, Leibniz, Stephenson, Watt, and dozens more. Apart from a glass case labelled "Oxford Geology" (focusing on William Smith, William Buckland and John Phillips), there is little attempt to explicitly identify Oxford contributions to natural science.

One notable historical occasion is marked. It was in 1860, at a meeting of the British Association convened here to celebrate the opening of the building, that the much publicised debate took place between Thomas Huxley (champion of Darwin and evolution) and Samuel Wilberforce (Bishop of Oxford, Dean of Christ Church). The Bishop asked Huxley whether it was through his grandfather or his grandmother that he was descended from an ape, or both. Huxley, white with anger, replied: "For myself, I would rather be descended from an ape than from a divine who employs authority to stifle truth". There is a poster display about the event and about evolution in general in two glass cases outside the door to what is now the lecture theatre on the upper floor. The debate itself took place in another room off the gallery — there is a plaque outside the door.

The University Museum is open Mon–Sat, 12–5.

4. Other Places of Interest

On the High Street, on a wall adjacent to University College, is a plaque marking the location of the former house of apothecary John Crosse, where Robert Boyle lived from 1655 to 1668 — Robert Hooke lived with him for much of that time. There were no separate work places for use as laboratories in those days; experiments were done right in this building and Boyle's Law for gaseous expansion was undoubtedly discovered here.

Halley's House.   Edmund Halley was Savilian professor of geometry from 1703 to 1742; the place where he had his home and observatory is at 7 New College Lane, close to the "Bridge of Sighs" that is part of Hertford College. It should be noted that Halley was 64 years old before he went to

The plaque on the house on New College Lane where the orbit of the famous comet was plotted.

Greenwich as Astronomer Royal; his plotting of the orbit of the famous comet and prediction of its return were done while he lived in Oxford. He knew he would no longer be alive when the great day came (in December 1758), but he expressed the hope that the world would remember that the prediction had been made by an Englishman, as indeed it has done. The event was an extraordinary public triumph of the scientific method and of Newton's mechanics in particular. In France it wiped out all vestige of belief in Descartes' views of the structure of matter, converting the French into fervent Newtonians.

Rhodes House.   A distinctive feature of Oxford life are the Rhodes scholars (created under the will of Cecil Rhodes), most of whom come from the United States. They may join any college of their choice, but they also have this building of their own, for special dinners and the like, on South Parks Road. Albert Einstein gave a lecture here in 1931 about his theory of relativity — the blackboard he used, with equations in his own handwriting, is preserved and on display in the Museum for the History of Science.

SLOUGH (Berkshire)

William Herschel (1738–1822) acquired a royal pension after his discovery of the planet Uranus, but one of the conditions was to move away from Bath to a place closer to Windsor Castle, so that King George III could pop over and peer through the telescope whenever he wanted. The chosen place was Slough and Herschel built a new forty-foot telescope for it, which turned out never to work as well as his earlier smaller instruments.

Herschel's home, Observatory House, was at the corner of Windsor Road and what is now called Herschel Street, where a concrete office building (ICL corporation) now stands. Outside, an abstract sculpture marks the spot, which claims to symbolise "the triangular structure of the telescope through which he reached up to encompass the infinite". Herschel was buried in St. Laurence's church (Upton), half a mile from his home. Close to the church is Herschel Park, from which one used to have an unobstructed view of Windsor Castle, about 2 miles away — today one can see only the raised roadbed of the M4 motorway.

The church is normally closed except for services. Phone: (01753)-529988 (vicarage).

This abstract sculpture commemorates William Herschel's telescope, through which, according to the inscription, he "reached up to encompass the infinite".

ST. ALBANS (Hertfordshire)

Francis Bacon (1561–1626) is a curious figure in the history of British science. He was a courtier and churchman and a philosopher of science, who in his writings claimed to espouse a rational scientific method, based primarily on experimental fact. But he himself, though a contemporary of many scientists who dealt in experimental fact (Gilbert, Kepler, Harvey, etc.) had no dealings with them and paid no attention to their published work. Nevertheless it became fashionable in later centuries to consider him a guiding spirit who influenced the researchers who came after him — whether he actually influenced anyone of note is questionable.

Bacon was knighted by James I, became his Lord Chancellor and later Viscount St. Albans. But he was dismissed from the chancellorship after a conviction for bribery and spent his last few years in retirement at Gorhambury (on the outskirts of St. Albans), writing philosophical works. After his death, his former secretary and fervent admirer (Thomas Meautys) erected a monument for him beside the altar in St. Michael's church — whether he is actually buried there is not entirely clear. The monument is unusual in showing its subject in an unheroic relaxed pose, leaning thoughtfully on his elbow. *Sic sedebat* ("thus he used to sit"), the inscription tells us.

Gorhambury. A lane leads from near St. Michael's church (beyond the Roman theatre) to Gorhambury, an 18th century house, seat of the present Earl of Verulam, where Bacon's library and other memorabilia are on view. Behind it are the ruins of the Tudor house in which he actually lived, which are maintained by English Heritage.

St. Michael's church and Gorhambury are at the heart of the most historic part of St. Albans, the site of the Roman city of Verulamium — directions to it are signposted on all major roads. Gorhambury is open to the public on Thursday afternoons only, May through September, but the mile-long lane to it is open to pedestrians at all times, except Saturdays in November, December and January. Phone tourist office, (01727)-864511, for full information. St. Michael's church is open for sightseers weekdays 2–3.30, Sat 2.30–5, Sun 2–6, May through September — otherwise only for services. Phone (01727)-8350137 (vicarage).

SUNNINGWELL (near Abingdon, Oxfordshire)

How far back can we go to find evidence for science at Oxford? To the very foundation of the university, it seems, for Roger Bacon (1220–1292), one of the university's earliest students, joined a Franciscan order at Oxford and (among other philosophical interests) had an understanding of optics and of the working of lenses that was far in advance of his time. The friary where he lived and worked was within what is now Oxford's Westgate Shopping Centre. The gate house to Folly Bridge, where he is said to have made astronomical observations, has also been demolished.

Fortunately, one memorial remains. The parish church of St. Leonard, in the sleepy little village of Sunningwell is another place where Bacon is said to have come to do experiments. A framed inscription on the porch commemorates the fact, with the aid of a little verse:

> Our little village church (they say) was old
> > When Roger Bacon loaned its gothic tower
> To furnish Lady Science with a bower,
> > Wherein to conjure Oxford clay to gold.

The last line alludes to Bacon's dabblings in alchemy, which was of course at that time a respectable occupation for a man of intellect. (The porch itself has a unique and much admired seven-sided design.)

Sunningwell is off the B4017, about 5 miles south of Oxford. The church was locked when we visited, but we obtained the keys from a lady at Yew Tree Cottage, just a few steps from the church. The formal rector has his principal appointment nearby at Radley. Phone (01235)-554739.

TRING (near Berkhamsted, Hertfordshire)

This is where Darwin's finches have come to rest — in an annex to the Zoological Museum in Tring! The museum was established in 1892 by Lionel Walter Rothschild, member of the banking family that at the time was one of the richest in the world, and it became through his care and support one of the finest collections of animal specimens in the world. Lord Rothschild had a passion for natural history and could afford to indulge it by purchasing existing collections, sending collectors all over the world to gather new specimens, and hiring expert resident curators to care for it all. He also provided for the public display of his specimens (requiring far more space than neat storage alone), though he was under no obligation to do so. Eventually he donated the museum and its contents to the Museum of Natural History, and it is today a satellite of that South Kensington institution. It is a thoroughly satisfying place to visit, with especially outstanding exhibits of birds and insects, lovingly preserved.

In 1971 the Ornithology Department of the Museum of Natural History was moved here from South Kensington, together with its research library and its priceless collection of bird skins, the ultimate avian taxonomy reference in the world. Its specimens (more than a million) are housed in 1500 cabinets on 3 floors, neatly arranged, expertly protected against insects and other agents of destruction. The jewel in the crown is a collection of skins of Darwin's finches from the Galapagos Islands, many of them from the _Beagle_ voyage itself, others brought in later by naturalists following in Darwin's footsteps — some were originally here in the Rothschild collection, were sold to the Natural History Museum in 1932, and have now returned "home".

The Ornithology Department's facilities are available to any scholar with a legitimate interest. Writing this book proved to be a valid justification and it was quite a thrill for us to have the curator unlock the appropriate drawer and pull out a couple of the Darwin skins — thick bill on one, thin bill on another, all evolved from a single ancestor — the very specimens which were so crucial in the foundation of Darwin's evolutionary theory.

The museum is open all year, 10–5 Mon–Sat, 2–5 Sun. Phone (01442)-824181. For permission to use the library or the skin collection, write to the Bird Group, Natural History Museum at Tring, Hertfordshire HP23 6AP.

Museum of Science and Industry. Birmingham's leading position in England's industrial midlands was originally based firmly on James Watt's steam engine. The Museum of Science and Industry on Newhall Street (in its James Watt Building) has one of the oldest working models extant, the Smethwick engine erected in 1779 to Watt's design. It is a monstrous engine, two floors high, which functioned to cycle water in and out of the locks of the adjacent canal. It still works and on the first and third Wednesday of every month they get up steam and have it running!

City Museum: The Lunar Society. The city is also commendably cognisant of its place in the more general history of science. Chamberlain Square, the local rendezvous for a bit of sun at lunchtime, has a Gothic fountain at its centre and James Watt and Josph Priestley on pedestals at the top of a flight of steps, overlooking the scene. Priestley is shown with a burning lens, directing the sun's rays onto a tube containing mercuric oxide, illustrating the method he used to produce pure oxygen. The town hall, council house, central library and the museum and art gallery line the square. The museum is principally dedicated to art, but it has a local history gallery on its lowest floor, which is well designed and informative and includes a panel on the Lunar Society, the little scientific discussion group founded by Erasmus Darwin, which met in the members' houses by the light of the moon so as to be able to find their way back home late at night. Pictures of twelve of its fourteen members are shown, including Darwin, Priestley and Watt and the latter's industrialist partner, Matthew Boulton.

Another panel describes the 1791 riots in which Priestley's house and laboratory were destroyed. The riots were provoked when a group of Birmingham dissenters arranged a dinner for 14 July 1791, the second anniversary of the storming of the Bastille in Paris, to celebrate and glorify the French Revolution. Printed announcements publicised the event — tickets were five shillings per head, including a bottle of wine. An angry mob formed. Claiming that church and king were threatened by the dissenters, the mob rioted for five days, attacking meeting houses (the dissenters' places of Sunday worship) and private dwellings. There is a vivid picture of the mob swarming over

Priestley's house on Fair Hill and smashing it to pieces. And a fine house it was, as shown by another picture of it dating from an earlier day. Priestley emigrated to America after this frightening experience and ended his days there (*see* CHEMISTRY p.42).

Both museums are open Mon–Sat 10–5, Sun 12.30–5. Phone (0121)-235-2834 for the Chamberlain Square Museum and (0121)-235-1661 for the Museum of Science and Industry.

University of Birmingham.   Mason College, founded 1875 in the centre of city, became the University of Birmingham in 1900 and then moved to its present site in Edgbaston. One of its early notable members was Birmingham native Francis W. Aston (1877–1945), who enrolled at Mason College in 1893 as a chemist and later returned to the University as a student in physics and subsequently for one term as lecturer. In 1910 he went to work with J.J. Thomson at the Cavendish Laboratory in Cambridge and on to eventual fame and fortune. Aston perfected the mass spectrometer — a device for separating atomic nuclei on the basis of their masses — and made it into the invaluable research tool it has been ever since. With it he discovered natural non-radioactive isotopes, the initial observation being with neon, which on analysis yielded two particles of relative mass 20 and 22, respectively. J.J. Thomson, as ever a sceptic, thought that the mass 22 particle must be a previously unknown hydride (NeH_2), but Aston proved him wrong and went on to show that many elements have two or more isotopes. Aston's work was crucial for the unravelling of the puzzle of how to account for measured atomic weights of the elements — how to reconcile them with the positions of the elements in the Periodic Table — and he was duly awarded the Nobel Prize for chemistry in 1922.

There appears to be no recognition of Aston in Birmingham or specifically at the university, the latter perhaps understandable because his residence there was brief. There is recognition of a later Nobel Prize winner, the organic chemist Norman Haworth (1887–1950), who was head of the university's chemistry department for 25 years. During this time he determined the molecular structure of vitamin C and followed that up by test tube synthesis from simple precursors — the first chemical synthesis of any vitamin. The chemistry building at the university is named after Haworth and the entrance foyer has an embossed medallion in his honour and a photograph of his research team.

Other memorials. Matthew Boulton, James Watt and William Murdock, another inventive emigré Scotsman, are shown together in a statue in front of the Registry Office on Broad Street. At first glance they appear to be choosing wallpaper for their offices, but it is presumably meant to represent a discussion of plans for a new engine.

The same three men have memorial statues or busts in St. Mary's Church in the suburb of Handsworth. This is where they all worked, where the Boulton–Watt factory was located. The factory was demolished long ago, but the house where Boulton lived is being restored and it is intended to open it to the public in the near future.

CHATSWORTH (near Bakewell, Derbyshire)

The Cavendish family (the Dukes of Devonshire) has been exceptionally prominent in the history of science in England. Henry Cavendish, the eccentric discoverer of hydrogen, was one of them, and his father (Charles) is in the record books as the inventor of the maximum–minimum thermometer. The mightiest deed was the creation of the Cavendish Laboratory in Cambridge by William Cavendish, the 7th Duke of Devonshire.

Henry Cavendish (1731–1810, *see* CHEMISTRY p.43) was an arch-eccentric. He denied himself any diversion or even the temptation for diversion — he gave his dinner orders, for example, by leaving notes on the hall table and his women servants were instructed to keep out of his sight on pain of dismissal. He left much of his work unpublished, in sealed packages in the family archives, which were not opened until the 1870s. They were then studied and edited by James Clerk Maxwell, first Cavendish professor at Cambridge, and revealed a more versatile scientist than had been realised before, who anticipated many modern theoretical concepts — the recognition of heat as molecular motion, for example.

(But uncommunicated thoughts don't do anyone much good.)

Chatsworth, the grandest of English country houses, has been the principal seat of the family since 1688 and has become one of England's most popular tourist attractions. The interior houses an unparalleled collection of art and sculpture, the celebrated gardens contain an orangery, a multitude of fountains, and a huge cascade. How big is the mansion? According to the current Duchess in a recent interview: "They say there are 175 rooms, but I've never counted them. It's fairly meaningless, because some are as big as squash courts and others smaller than laundry baskets." Henry Cavendish's surviving manuscripts are held in one of the rooms and can be seen by appointment.

Chatsworth is open April to October, daily 11–4.30. Phone (01246)-582204.

CHESTERFIELD (Derbyshire)

George Stephenson, the railway pioneer, spent the last ten years of his life in Chesterfield, having originally moved here to supervise the building of the North Midland railway line from Derby to Leeds. He resided at Tapton House, about a mile east of the town and he was buried in Holy Trinity Church in the town centre. Just across the street from Chesterfield's most famous landmark, the church with the crooked spire, the town erected the Stephenson Memorial Hall in 1879, housing a theatre and an art gallery.

Holy Trinity Church (Newbold Road) is kept closed except for services, but anyone interested in seeing Stephenson's tomb can readily obtain the keys from the rectory at 31 Newbold Road. Phone (01246)-232048. Tapton House was recently used as a school, but was unoccupied when we visited in 1994.

A more recent connection with science is provided by Robert Robinson (1886–1975), one of Britain's most prominent organic chemists, winner of the Nobel prize in chemistry in 1946 for working out the detailed atomic structures of strychnine and other complex molecules. Robinson was born (the son of a local manufacturer) at Rufford Farm, 3 miles west of Chesterfield on the A619; he was a restless soul and worked at several universities, ultimately at Oxford.

DERBY (Derbyshire)

The city of Derby has never itself been intimately involved with basic scientific activities, but it has had native sons, residents and patrons, who have made major contributions. It is an attractive city to visit, with a well-planned pedestrian area at its centre.

Foremost among the natives is John Flamsteed (1646–1719), the very first Astronomer Royal at the Greenwich Observatory in London. He was born in Denby (6 miles to the north) and was educated at the historic Derby grammar school, first given its charter by Mary Tudor in 1554. The building has been restored and is now part of the Derby Heritage Centre, which promotes local history and the sale of local arts and crafts. A later renowned native was Herbert Spencer (1820–1903), social theorist and psychologist, but also an early advocate of evolution, albeit of the Lamarckian (adaptive) variety — he coined the phrase "survival of the fittest" in 1852, some years before publication of Charles Darwin's *Origin of Species*. The most prominent resident was Charles Darwin's grandfather, Erasmus Darwin, who lived and practised medicine here from 1781 until his death, and who founded the Derby Philosophical Society (modelled on the Birmingham Lunar Society) in 1783. Darwin and Spencer are remembered by plaques at the corners of the bridge across the Derwent River on Derwent Street — both lived nearby, Darwin on Full Street and Spencer on Exeter Row. (Literature buffs may be interested to know that Herbert Spencer was unmarried, but had an affair with Marian Evans, also known as George Eliot).

Among Derby's patrons we have the Cavendish family from the great Derbyshire estate of Chatsworth. Until the 19th century the family members were buried in the parish church of All Saints (now Derby Cathedral) in vaults lying immediately below the "Cavendish Area" on the south side of the altar. This is a splendidly decorated part of the church and displays a selection of coffin plates fetched up from below, including one for Henry Cavendish, the outstanding scientist in the family (*see* CHATSWORTH, above), who died in 1810. The dominant feature of the area is the grandiose tomb of Bess of Hardwick (1520–1607), one of the founders of the dynasty and builder of Chatsworth and other grand country homes. Some insight into the character of this remarkable woman is gained from the fact that the monument and laudatory inscription were put up by her own instructions during her lifetime.

Derby Heritage Centre is at St. Peter's Churchyard in the pedestrian area: it is open daily all year. Derby has a rich industrial heritage: its silk mill (1717) is said to have been the first true factory in England and has now become the Derby Industrial Museum. Phone (01332)-299321 (Heritage Centre) or (01332)-255802 (tourist office).

IRONBRIDGE (near Telford, Shropshire)

Ironbridge is unique, a widely promoted tourist centre with good facilities, but devoted entirely to industrial technology and the science behind it, with no frivolus distractions. More specifically, it gives us the story of iron and steel. This is the place where commercial ironmaking began and its visual centrepiece is the grand iron bridge across a gorge of the River Severn, the first of its kind, built in 1779.

There are several museums and other designated sites of interest, where you cannot fail to learn (if you did not already know) many things: that iron exists in the rocks as oxides, which must be reduced with carbon to make the metal; the distinction between wrought iron and cast iron and steel; the second use of carbon, to combine chemically with the metal, and how that affects its properties. The sites are spread along a 4-mile stretch of the river and most of a day should be set aside for even a casual visit. They include the original Darby furnace at Coalbrookdale, where Abraham Darby (in 1709) pioneered the technique of smelting iron with coke instead of charcoal. Nearby is the house where Darby lived. A half mile away, by the river, is a general museum (and visitor centre), housed in the shed where the iron product used to be stored prior to shipment by river

Iron bridge across the Severn, built in 1779.

barge. At the opposite end of the area is Blists Hill Open Air Museum, partly a reconstructed Victorian "town", but also the place where you can actually see iron being cast by the old methods. In between we have the iron bridge itself, and a cave where you can don a hard hat and go underground yourself, and other attractions.

Ironbridge lies between the A4169 and A442 and is copiously signposted. It is open daily all year (except Christmas), 10–5. Some sites may be closed at times in winter and actual casting at Blists Hill can only be seen on certain days. Phone (01952)-433522 or (weekends) 432166. For direct contact with Blists Hill, phone (01952)-583003.

JODRELL BANK (near Macclesfield, Cheshire)

In 1931 an American engineer, investigating atmospheric interference with radio signals (so-called "static"), discovered that the source of some of the interference seemed to be moving across the sky, as if it were coming from a star. Little effort was made to exploit this phenomenon as a possible scientific tool until after the war, when Bernard Lovell, professor at the University of Manchester and fresh out of a term of duty in radar engineering, decided to tackle the task. An immediate prime need was to find a site free from artificial radio signals (the counterpart of the 1931 "interference" problem) and its solution was found at Jodrell Bank, already used as an experiment station by the university's botany department. Thus radio astronomy was born — looking at radiation from the sky in the radio-frequency range of the electromagnetic spectrum rather than the narrow band of visible light that the human eye can see. It has become an almost equal partner with light astronomy in the investigation of the universe, in the forefront of research on everything from comets nearby to faint signals from the most distant galaxies.

The giant Lovell telescope we see here is the second largest such device in the world — not surprisingly, it looks

Black holes, anyone? Jodrell Bank is Britain's prime centre for radioastronomy.

basically like a satellite TV dish, for the principle of focussing radio waves is the same in both. Adjacent to the telescope we have a well-designed "science centre", in which the related basic physics is explained — the electromagnetic spectrum from gamma rays to radio waves, our present notions of our galaxy and the universe and their origins, etc. There are animated figures of "Newton" and "Einstein" which speak to you — a gauche and even somewhat offensive gimmick. But stressing that Newton's mechanics and Einstein's relativity are essential to an understanding of what is going on here is in itself commendable, for it would have been easy to forget the past and promote the project entirely in terms of today and (especially) tomorrow — the wonders yet to be discovered, black holes, life on other planets, and the like.

Jodrell Bank Science Centre is off the A535, between Holmes Chapel and Chelford, and clearly signposted. It is open daily, 10.30–5.30, from the middle of March till the end of October — the rest of the year weekends only, 11–4.30.

LICHFIELD (Staffordshire)

Lichfield is a pleasant place to visit, with a rather unusual cathedral, a fine park near the town centre, and many monuments to Samuel Johnson, who was born here in 1709. (And even a statue to James Boswell, the publiciser of Johnson's life and deeds.) Perhaps surprisingly, the town also makes quite a little fuss over Erasmus Darwin, who had his medical practice here for twenty-five years, but devoted himself with equal energy to the affairs of Birmingham's Lunar Society and to his own scientific work and publications. His greatest contemporary influence was through his championing of Linnaeus's system for the classification of plants, which is based on the sexual parts of the flowers. The system was still being criticised at the time and Erasmus Darwin consciously set out to promote it, driven by the conviction that sexual reproduction is the basic foundation for all of life, obviously so for animals, but equally so (even if less obviously) for plants. He popularised the idea, following Linnaeus in his manner of stressing sexuality — here is a typical extract from his poetic book, *The Loves of the*

Plants, giving Darwin's description of *Lychnis,* the scarlet campion of the meadows around him:

> Each wanton beauty, trick'd in all her grace,
>   Shakes the bright dew-drops from her blushing face;
> In gay undress displays her rival charms,
>   And calls her wondering lovers to her arms.

At the foot of the page, in fine print, the botanical facts are set forth in more conventional prose — but the reader can see why Darwin and his friends are often called "eccentrics", for none of this would have gone over well in Oxford or Cambridge.

Darwin's imposing residence at the edge of Cathedral Close still stands and is marked by a plaque. The cathedral, which is crowded with many monuments, has a brass plate in Darwin's memory in the aisle of the south choir. Lichfield's Science and Engineering Society sponsors an Erasmus Darwin Memorial Lecture and other local groups have laid out the *Darwin Walk,* a ten-mile trail around the town which is supposed to recreate the flora and fauna of two hundred years ago. (Different towns have different styles: Shrewsbury, Charles Darwin's birthplace, is much more niggardly in the celebration of the Darwin name than Lichfield, although Erasmus was a lesser figure than his grandson and only a transient resident here.)

Lichfield is 16 miles north of Birmingham.

MAER (near Stoke-on-Trent, Staffordshire)

Maer Hall, in the village of the same name about ten miles southwest of Stoke, is a historical site for any scientifically oriented visitor because of its intimate connection with Charles Darwin — the fact that it can only be viewed at some distance from the outside is no drawback, for it is the ambience of the setting that fascinates.

Maer Hall was the home of Josiah Wedgwood II, the elder Josiah's son and Charles Darwin's father-in-law. The Hall (which has passed through several owners) and the gorgeous estate on which it sits have changed little since the Wedg-wood days, except that much of the surrounding land no longer belongs to the owners of the manor. Seen from the chuchyard on the hill across the street it gives an impression of enormous wealth and good living, testimony to the highly privileged segment of society from which Charles Darwin derived. Charles himself writes about his visits to Maer as a teenager. He loved the shooting and the family atmosphere:

> In the summer the whole family used often to sit on the steps of the old portico, with the flower garden in front, and with the steep wooded bank, opposite to the house, reflected in the lake, with here and there a fish rising or a water-bird paddling about. Nothing has left a more vivid picture on my mind than those evenings at Maer.

Charles presumably first became acquainted with his future wife at this time — she was Emma Wedgwood, his mother's cousin. They were married about ten years later in 1839 (after the voyage of the *Beagle*) in the Parish Church of St. Peter, which stands connected to the Hall by a stone bridge across the road. The marriage certificate is reproduced in a pamphlet that can be purchased in the church.

MIDDLETON HALL (Warwickshire, near Birmingham)

Middleton Hall is the birthplace and home of Francis Willughby (1635–1672), the comrade of naturalist John Ray (*see* BOTANY p.12) in his travels and in the intent to prepare a comprehensive flora and fauna — he was also Ray's bene-factor, for Ray was poor and could not have financed the project on his own. Willughby's early death might have ended the project, but he left his friend a perpetual allowance and Willughby's mother encouraged Ray to stay on at the Hall and to continue the work there, which he did, editing and publishing the fauna part of the project, which Willughby had partially completed. This arrangement con-tinued for several years, until the elder Mrs. Willughby died — Francis's wife, who had never liked Ray, then made him leave and he returned to his own birthplace, Black Notley in Essex.

The Hall remained in the Willughby family until 1924, when it was sold (to pay inheritance taxes) to a company that set up quarries on the land, but left the buildings to decay. The estate is now leased to an enthusiastic volunteer group, who have restored part of the property and are working on the rest. The principal building is Georgian (erected about 1710). One room has been set aside in it as a "history room" and is full of factual information about the house and the people who lived there. Pride of place goes to Francis Willughby and John Ray and a facsimile edition of Willughby's _Ornithologia_ is on display — a treatise written here by Ray after Willughby's death and based partly on his notes. A small annex building (still far from restoration) has been identified as the place where Ray may have lived; the building where he did his work no longer exists. Some outbuildings have been converted into a craft centre — undoubtedly the chief attraction for most of those who visit here.

Middleton Hall is on the A4091, 12 miles northeast of Birmingham, 4 miles south of Tamworth; the access road is clearly signposted. The site and craft centre are open Easter to October, Tue, Thur, Fri, Sat, Sun, 11–5.30. The interior of the Hall is only open on Sundays and holidays, 2–5.30, but there may be people on hand to show it at other times when the site itself is open. Phone (01827)-283095.

NOTTINGHAM (Nottinghamshire)

Nottingham's most eminent scientific citizen is the mathematician George Green (1793–1841), whose name is a household word in the vocabulary of theoretical physicists and applied mathematicians, chiefly for the creation of a theorem that connects surface and volume integrals in such a way that physical quantities for the _surface_ of any object (or any part of space) can be directly related to conjugate quantities for the _volume_ contained within that surface — an indispensible tool for many kinds of theoretical analysis.

The extraordinary fact about Green is that he was a miller, employed in his father's milling and baking business. His

Green's Mill, monument to a great mathematician.

education was minimal. There is conjecture about the source of his self-acquired mathematical knowledge, that the stimulation came from joining a recently created local library at the age of 30, but concrete facts are sparse until five years later (1828), when he published his most important work, a mathematical analysis of theories of electricity and magnetism — publication was by private subscription, with only 52 subscribers! Green's theorem, as it is called, was contained within this work.   After the death of his father and sale of the mill, Green (now aged 40) was able to enter Caius College, Cambridge, for a degree course in mathematics, and he subsequently became a fellow of the college and published several more mathematical works. But his original work remained unknown and its value was not appreciated until William Thomson (later Lord Kelvin) discovered it in 1845: he had it republished a few years later with an ethusiastic introduction by himself. In the meantime Green had returned to Nottingham and died there of influenza in 1841. Neither he nor any of his fellow townsmen could have imagined the esteem in which he is held today.

Green's mill was restored in 1986 as a public monument and a tiny one-room museum was added beside it. The museum has some simple hands-on experiments in optics for youngsters, and Green's formula for the relation between volume and surface integrals is used as decorative motif throughout the room, but without definition of what the symbols mean or even what an integral is — not surprising, perhaps, in an exhibit designed for small children. There is a modern sculpture in the courtyard, entitled "Green's Theorem Sculpture", but its relation to the subject was not obvious.

Green is buried in a family grave close to the mill, in the grounds of St. Stephen's church — the stone is inscribed to the memory of his father and mother, their son (i.e. George Green) and the latter's son, who died at the early age of twenty.

Green's Mill is in the suburb of Sneiton (east side of Nottingham), at the corner of Sneiton Road and Windmill Lane. It is open all year Wed–Sun, 10–5. Phone (0115)-950-3635. The churchyard is on the opposite side of Sneiton Road; the tomb is in the corner closest to the "Fox" pub.

OSWESTRY (Shropshire)

Edward Lhuyd (1660–1709), a notable Welsh scholar among Britain's "early lights", came from a family that had lived in the border area for hundreds of years. He was a naturalist, close friend of John Ray, and (among other works) published an early catalogue of British fossils. His most penetrating work was probably as a student of Celtic languages: the distinguished linguist John Rhys described him as "the greatest Celtic philologist the world has ever seen". He travelled far and wide in pursuit of his interests, collecting both specimens and "words": a memorable trip to Cornwall resulted in the only scholarly record of the

Cornish language, then still spoken in parts of the county.

Lhuyd attended Jesus College in Oxford and was the first keeper of the Ashmolean Museum there when it was opened in 1683 — there is a tablet in his memory at the College. John Rhys has called upon Oswestrians to erect a statue to celebrate this local genius. It has not been done, but the half-timbered family home (built in 1604) survives, prominently labelled "Llwyd Mansion" — an alternative spelling of the name. It displays the family coat of arms and looks almost the same today as in sketches made at the time it was built. Edward, it should be noted, was illegitimate and he was not born in this building, but it was most likely his home for many years when he was a student at Oswestry grammar school.

Oswestry is a market town with an attractive hilltop centre for municipal and district offices. Llwyd Mansion is at the corner of Bailey and Cross Streets; it houses a dry cleaner and other shops; its striking external appearance is its only distinguishing feature.

SHREWSBURY (Shropshire)

Shrewsbury, as seen by the visitor, is a constricted town, with most of the old parts squeezed into a loop of the River Severn. Charles Darwin was born here in 1809 and there is a bronze statue of him in this inner area, near the Castle, in front of the town library. The library used to house Shrewsbury School, where Darwin received his education to age 16, at which time he left to study medicine in Edinburgh. (Earlier famous students of the school included Judge Jeffreys, the "hanging judge" of the reign of James II.) Outside the Severn loop across Welsh Bridge is Darwin House ("The Mount"), where Charles was born and where his father practised medicine — it is now an office building, with the gates closed outside office hours. Darwin's father and mother are buried at Montford, about five miles to the west.

Shrewsbury's pride in its famous son is on the whole rather muted. Do not be misled by signs pointing to Charles Darwin Centre — that's a small shopping centre off the main shopping street.

Charles Darwin's statue, in front of the building where he went to school.

SILURIAN ROCKS (south of Shrewsbury, Shropshire)

Roderick Murchison, Scottish-born geologist, was in competition with Adam Sedgwick in the search for the oldest fossil-bearing rocks in Britain (_see_ GEOLOGY & PALAEONTOLOGY p.33). The area bordering on Wales in Shropshire was his favoured hunting ground and his explorations here led to the definition of the Silurian era in geology, named after the Welsh tribe that once dominated the land. We cannot conveniently travel in Murchison's exact footsteps, but we can identify the mountain areas where he used his hammer and chisel and see the steep and bare escarpments that create a high vertical sampling range.

Long Mynd (Church Stretton). Church Stretton is 12 miles south of Shrewsbury on the A49. Long Mynd ("Mynd" is Welsh for "mountain") lies to the west on a steep road that perfectly illustrates the persistent feature of the Silurian area: precipitous escarpments on the north side and much gentler grassy slopes to the south. The road (mostly single track) is signposted "Burway road" and actually cuts right up the escarpment. There is a car park about a mile out of Church Stretton, where you can look back on other heights that feature in Murchison's description, notably Caer Caradoc, just two miles away, which defines one of Murchison's chronological subdivisions of the Silurian rocks. (The total sequence is Ludlow, Wenlock, Caradoc, Llandeilo.)

Wenlock Edge. The B4371 from Church Stretton to Much Wenlock (12 miles) runs for much of its length along the ridge of Wenlock Edge, an escarpment which has become internationally famous for its rock exposures and limestone flora. Especially interesting areas belong to the National Trust, which publishes a pamphlet about what there is to see and maintains signposted car parks for access to trails along the ridge. The town museum in Much Wenlock has a small geological exhibit, which provides minimal information, such as the fact that the rocks we see were laid down 400 million years ago, when the land was still submerged beneath shallow seas. More important, it too is a source for inexpensive booklets, e.g., "Wenlock Edge — geology teaching trail", published by the Nature Conservancy Council.

The museum in Much Wenlock is in the town square. It is open 1 April–30 September, Mon–Sat (also Sun in July and August) 10.30–5. Phone (01952)-727773. Church Stretton has an information centre with knowledgeable staff, open Easter to end of September, in the same building as the public library. The rest of the year some of the same personnel work in the library itself. Phone (01694)-723133.

STOKE-ON-TRENT (Staffordshire)

This is the land of Wedgwood, where the tourist information

centre will sell you (if you wish it) a car cassette to guide you on the Wedgwood Trail around the city and environs, to learn about the life and times of Josiah Wedgwood (1730–1795).

Wedgwood is of course known to all the world as a pottery designer and manufacturer, founder of the Staffordshire factories that still bear his name. His inclusion in the scientific community is less well known, but he was actually an active member of Birmingham's Lunar Society and generously supplied fellow-Lunarite Joseph Priestley with crucibles and other ceramic ware for his research. He himself took a scientific approach to pottery-making and invented a pyrometer for measuring the high temperatures in firing ovens, which earned him membership in the Royal Society. Even without such tangible contributions he would have to be included in our book for a more personal gift, as progenitor of Charles Darwin — his daughter Suzannah was Charles Darwin's mother, he and fellow Lunarite Erasmus Darwin share the grandfather honours and no-one can tell whose genes were the more important. (Curiously, when Charles himself came to take a wife, he married his first cousin Emma, also a grandchild of Josiah Wedgwood.)

Pottery is still Stoke's bread and butter and the city's tourism is centred on its history and present activities. You can see Burslem, the "Mother of Potteries", on the northern outskirts of the city, where the first Wedgwood factories were located, and Etruria, where manufacturing was done from 1769 to 1950 — there is an industrial museum there and the Wedgwood family home, now skilfully incorporated into the Moat House Hotel. The City Museum has additional historical material.

Barlaston. The present factory (and the most elaborate tourist facility) is on a picturesque estate near Barlaston, 5 miles to the south. The Wedgwood Visitor Centre there has a good museum, devoted mostly to pottery making and design — the ornamental black or blue vases and urns of Josiah's days are truly delightful. The first room of the museum stresses Josiah's scientific interests and his friendship with Joseph Priestley and others, but there is only one small showcase with actual samples of Priestley's ceramic ware. Understandably, the shop is the most prominent part of the centre — no visitor (it is hoped) will leave without a piece or two of the famous pottery in his shopping bag.

The Wedgwood Visitor Centre is open Mon–Fri 9–5, Sat and Sun 10–5, but closed Sun from October to Easter. Phone (01782)-204141 or 204218. Contact the tourist office for scheduled factory visits in Stoke. Phone (01782)-284600.

WARRINGTON (Cheshire)

The dissenting academies which proliferated in England in the middle of the 18th century have been described (*Oxford History of Britain*) as providing "progressive education designed to fit the sons of the middling sort to staff the professions and the world of business". In other words, they were the making of middle class Britain. Their teaching was done in English — Latin not required; they were often associated with noncomformist chapels, but church membership was not a requirement for participation, as it had been (since the restoration of the monarchy) in Oxford and Cambridge.

Warrington Academy was the most successful of these institutions. Joseph Priestley was among the tutors (1761–1767), but not as a chemist — in fact it was here that he attended his first lecture in chemistry and acquired his lasting interest in the subject. Robert Malthus, whose name would become a household word for social reformers and scientists alike, was a student around 1780. He was born in Surrey (*see* ALBURY, Surrey p.102) and returned there after his schooling. His parents sending him this far north for his education is eloquent testimony to the Academy's quality and far-reaching influence.

Warrington today is a particularly pleasant town. The original building of the Academy (used from 1757 to 1762) has been preserved and restored; it stands at the foot of Bridge Street (bridge across the Mersey), now part of the offices where the *Warrington Guardian* is published. There is a statue of Oliver Cromwell beside the building and a plaque to commemorate the Academy. In its own time, the Academy quickly outgrew its original site and moved to larger quarters nearby, now remembered only by the street name, Academy Street.

WIDNES (Cheshire)

We find ourselves here at the heart of Britain's chemical industry. It began with the alkali industry, which took root in the middle of the 19th century in the area around Widnes and Runcorn — a natural location, with salt, lime and coal readily available and transport provided by the river Mersey and the port of Liverpool. Its products (starting from salt and sulphuric acid) included alkali for making soap from animal fats, hydrochloric acid for bleaching cotton, sodium sulphate for glass manufacture, bicarbonate of soda for baking and medicine. There were 24 chemical works in Widnes by the 1870s. They evolved rapidly in response to market-driven advances in methodology and were forced into other kinds of change by community-driven outcries against stench and pollution — we can trace a direct line through mergers and acquisitions to the chemical giants of today (ICI, Unilever), which are still centred in the Merseyside area.

The Catalyst Museum in Widnes is designed to promote public understanding of this history and modern activities of the chemical industry. It is built on the site of the former "Gossage's Soap Works" and has an exceptionally successful educational design. Whereas the technical exhibits of other museums tend to display complete machines, impressive (even intimidating) in their complexity, we are here given the individual components — a good display of chemical engineering tools such as a gas scrubbing tower and a fluidised bed of solid particles. There are nice hands-on experiments showing crystals melting and reforming, hydrolysis of water, etc. Computers are used to explain each apparatus, giving individual visitors broad scope in how much they might want to learn. Poster displays summarise the careers of some of the people involved, e.g. William Gossage (1799–1877) and William Lever (1854–1925). The museum is still in process of development; an extension of the building and new exhibits are projected for early 1995.

The museum is on Mersey Road, just off the approach to the Runcorn–Widnes bridge across the Mersey — signposts show the way. It is open Tue–Sun, 10–5. Phone (0151)-420-1121. If coming by rail, note that Runcorn, not Widnes, is the closest station.

6
Eastern England

BLACK NOTLEY (near Braintree, Essex)

This is a tranquil village in the midst of farming country. The naturalist John Ray (*see* BOTANY p.12) was born here in 1627, the son of a poor blacksmith. He first set out to be a theologian, but the wonder of the flowers of the countryside drove him to botany instead. He became an academic, close to home in Cambridge. He explored much of Britain and the European continent with his friend Francis Willughby, and together they planned to write a comprehensive flora and fauna — all the species that they knew of. After Willughby's early death in 1672, Ray remained for a while in Willughby's Warwickshire manor house, editing manuscripts his friend had left unfinished, but eventually he came back to Black Notley and it was here that most of his published work was actually written. A commemorative plaque on a high wooden post was installed in the village centre in 1986, the tercentenary of the publication of Ray's *Historia Plantarum*. The house where Ray was born, his father's forge, survived until fairly recently, but was destroyed by fire and no trace remains.

Ray is buried in the graveyard of the church of St. Peter and Paul, which stands well outside the modern village, opposite an immense farmyard with a large 16th century wooden barn. Ray's grave is identified by a tall tombstone

next to the church entrance. The original Latin inscriptions on its sides are badly worn and difficult to read, but a modern inscription and plaque were put on by the Ray Society in 1984. An English translation of the original was placed inside the church at the same time.

Braintree. The Braintree District Museum is planning to devote a room to Ray and his work. A statue of Ray, now in the town centre, will be moved to the museum forecourt at the same time.

Black Notley is 2 miles south of Braintree. Phone (Braintree tourist information) (01376)-550066.

CAMBRIDGE (Cambridgeshire)

Where else can we find so much of the history of British science? Isaac Newton came here in 1661 and stayed for 35 years, Charles Darwin found his vocation here after years of indecision, the Cavendish Laboratory (founded 1871) brought unrivalled eminence in physics, Watson and Crick (in 1953) discovered the molecular basis of inheritance. "Oxford for arts, Cambridge for science," says an old adage, which is of course an oversimplification, but there is a germ of truth in it — Oxford cannot match these pinnacles of Cantabrian science.

Entries for Cambridge are grouped as follows:

1. Cambridge Colleges
2. Science Laboratories
3. Museums
4. Other Places of Interest

---

1. Cambridge Colleges

Christ's College. Christ's is a college away from the river, near the city centre. It is the college attended by Charles

Darwin, with parental instructions to prepare himself for a career as clergyman. Darwin, however, had little enthusiasm for the prescribed studies — by his own admission, he went to very few lectures and took more interest in his beetle collection than in his assigned reading. More important, it was in Cambridge that Darwin formed his decisive friendship with the botanist John Stevens Henslow, who subsequently recommended him for the position of unpaid naturalist on the Royal Navy's surveying ship, _H.M.S. Beagle_ — the 5-year journey that followed was the source of all the factual data that led to Darwin's evolutionary theory. Darwin is commemorated at the college by a bust and by a little piece of garden, known as the "Darwin Shrine", set in an inconspicuous corner behind the Master's Garden.

Another former member is the writer, C.P. Snow, who was a physics student while at the college. Much of his writing attempts to mirror collegiate life and scientific endeavour — his novel _The Search_ is based on real characters and events in the earliest days of molecular biology.

Darwin College. Three of Charles Darwin's sons — the ones who had a scientific career — eventually settled in Cambridge. Darwin College, across the river on Silver Street, was built around the former home of George Darwin (1845-1912), who was Plumian professor of astronomy for nearly thirty years and worked mostly on terrestial applications of his field, notably a definitive theory of tides that allowed for the lag induced by internal friction (viscosity) of sea water. The other two sons with a scientific bent, George's brothers, were Francis, who was a botanist, and Horace, the youngest, who ran something called "The Shop", considered by the family at the time to be a rather dubious enterprise. It eventually turned out to be profitable and became the foundation for today's Cambridge Scientific Instrument Company.

The George Darwin home has a certain notoriety because it is the centre of action of _Period Piece_, an amusing and irreverent book by Gwen Raverat, George's daughter, which is all about growing up in Cambridge in the 1880s. Darwin College (created in 1964) is an exclusively postgraduate college jointly managed by Trinity, Caius and St. Johns — it is quite different in function from the traditional undergraduate college. The old part of the building still looks much like the sketches in Gwen Raverat's book and the traffic outside is just as heavy. A large painting of Charles Darwin hangs upstairs, outside the master's office.

Gonville and Caius. This college, between King's and Trinity, was originally founded as Gonville Hall in 1348, but did not really prosper until it was refounded 200 years later by John Caius (1510–1573) — it is usually referred to as Caius College, and the visitor (to avoid being held in contempt) should know that the word is pronounced "keys". Caius was a Gonville graduate and had an M.D. degree from the University of Padua, where he studied with the great anatomist Andreas Vesalius — actually lived in his house for 8 months. He had enormous influence on English medical practice, fighting for rigorous licensing, laws against quackery, etc., and himself made modest contributions to medical scientific knowledge. Caius Court, small and intimate, quite different from the large expanses of adjacent Trinity College, was added by him and is in a sense his memorial. Its showpiece is the "Gate of Honour" on the south side, through which students to this day leave the college when they graduate, having entered years earlier through the "Gate of Humility". Caius's tomb is in the chapel, the elaborate stone coffin recessed into the south wall; it bears the simple inscription, *Fui Caius*; in English "I was Caius".

The college's most famous former member is William Harvey, who followed in Caius's footsteps by going to Padua for his formal medical education, and subsequently made the monumental discovery of the circulation of blood. Other graduates include Thomas Gresham and (more recently) another physiologist, Charles Sherrington. A stained glass window in the Hall commemorates the mathematician George Green, who entered the college for a degree course in 1833 (at age 40) *after* he had done the work for which he is most famous (*see* NOTTINGHAM p.181). Finally, mention should be made of a most unusual architectural feature: a frieze on the front of the college, bearing carved faces of notable graduates, including Caius, Harvey, Gresham and William Wollaston.

King's College.   Is economics to be classified a science? If so, then Maynard Keynes — proponent of the theory for government interference, many elements of which are now almost universally followed — must surely be considered one of its great pioneers. Keynes was a member of King's, first as student and after 1908 as fellow. In 1919 he became a bursar and in that capacity inaugurated a more adventurous financial policy that enormously increased the college's wealth, which we can take as proof that he had practical as well as theoretical skill as an economist. Today there is a building named after Keynes and a hall for public lectures and other meetings, with a separate entrance on King's Lane. A bust of Keynes stands within.

King's College is of course most celebrated for its magnificent Chapel and for the choir that sings there, a "must" for every visitor to Cambridge. Side chapels are used as an informative museum, limited however to stained glass, royal patronage, and other matters directly related to the Chapel alone.

Queens' College.   Erasmus of Rotterdam (1467–1536), humanist, satirist (author of *In Praise of Folly*) and religious reformer, is the celebrity here. Does he rate mention in a book dealing with science? The superficial answer would be no, but it is difficult today to judge the relative historical influence of a man like Erasmus — the most persuasive of those who argued broadly for an open mind and against authoritarian teaching — as compared, for example, to his contemporary, Thomas Linacre (*see* OXFORD p.155), who held a similar philosophy, but explicitly extended his activities into medical science. Among the sights to see at the college are the wooden "Mathematical Bridge" across the river, so-called because it is said to have been constructed on mathematical principles so as not to require any nails to hold it together — many subsequent rebuildings suggest that the original principles may not have been quite sound. There is an Erasmus Tower, above the rooms where the great scholar studied, and a modern building opposite the chapel is also named after him.

Two successive queens were involved in the foundation of the college, which accounts for the plural name: Queens' rather than Queen's.

The "Mathematical Bridge" at Queens' College.

Trinity College.   Trinity College must head the list (for readers of this book) of all the splendid colleges whose green lawns back onto the river Cam. Its Great Court, entered from Trinity Street through the main portal, is the largest of its kind, with an ornate central fountain, an ancient sundial, the college chapel on the north side, the colourful great gate, and the clocktower adjacent to it — surely one of the most impressive sights in the town. The clock, incidentally, strikes the hours twice in succession, first in deep bass and then in a lighter tone.

The appearance of the college has changed remarkably little since Newton came here as a student in 1661 for what would prove to be his residence for more than thirty years. His rooms (for about the last half of his stay) were in front, above the porter's lodge, between the main portal and the chapel — he had his own steps (now gone) leading down to a little garden below and on to Trinity Street. The chapel has a rather large vestibule, which today contains a statue of Newton alongside other distinguished Trinity Fellows: Francis Bacon (long before Newton); Isaac Barrow, Newton's predecessor as Lucasian Professor, who was the first to recognise Newton's genius and nominated him as successor when he himself resigned; William Whewell, the research as well as for hands-on experiments by students (as distinct from "demonstrations" performed by a lecturer).

James Clerk Maxwell (*see* ELECTRICITY & MAGNETISM p.19) was the first Cavendish Professor and he personally supervised the construction of the laboratory and the purchase of equipment. In effect, he set the tone for the future, making the Cavendish one of the best equipped physics laboratories in the world. Lord Rayleigh came next after Maxwell's early death, followed by J.J. Thomson, Ernest Rutherford and Lawrence Bragg, usherers-in of subatomic physcis and x-ray analysis, all of them among the early Nobel laureates in physics. Rayleigh's award was in 1904, given for the discovery of argon, Thomson's in 1906 was for the discovery of the electron. Rutherford and Bragg had received their awards before they came to Cambridge, but continued with undiminshed brilliance at the Cavendish and here trained a host of creative successors.

The Cavendish is also the place where Watson and Crick ushered in the modern era of molecular biology in 1953 with their elucidation of the structure of DNA — the celebrated double helix. This was no accident but the result of the conviction of many individuals (especially Lawrence Bragg, the then director of the laboratory) that a blending of the methods of structural physics with biological insight would produce proud results . This must have been coupled with indulgent flexibility by the managers of Britain's relatively meagre post-war resources, for there cannot initially have been much hope of success for such a far-out project. The detailed story of the actual progress on this problem, and related discoveries that were made in Cambridge about the same time, has a uniquely English flavour — an important factor was obligatory afternoon tea in the common room and the exchange of ideas that went along with it. (Two successful popular books have been written about the discovery by the protagonists themselves: *What Mad Pursuits* by Francis Crick and *The Double Helix* by James Watson. Read them both to get two very different pictures of "genius" in the laboratory.)

The Cavendish Laboratory was on Free School Lane, a narrow street just a couple of hundred yards from King's College. The site was abandoned for more modern quarters in 1974 and the building is now the home of a miscellany of

The old Cavendish Laboratory on Free School Lane.

minor university acitivities. A plaque reminds us of its days of greatness and the Maxwell lecture theatre remains — it is still the locale for undergraduate physics lectures.

New Cavendish Laboratory. The new Cavendish Laboratory is about a mile and a half to the west, on Madingley Road, on a lovely site, with its own pond and adjacent to open fields. Its three principal buildings are named after Bragg, Rutherford and Mott and its library is named after Rayleigh. Current research is particularly strong in radio astronomy and superconductivity.

A unique and fascinating feature of the laboratory is a small museum on the first floor of the Bragg building. It contains experimental equipment designed and used by the illustrious holders of the Cavendish chair, the actual apparatus for the classic experiments — done in the *old* Cavendish of course — that ushered in the era of subatomic physics. In today's climate of "big money" physics (super-colliders and the like) we cannot help but marvel at the simplicity of the apparatus we see here: the Crookes tube used by J.J. Thomson to "discover" the electron; the nuclear disintegration chamber which Rutherford used to bombard elements with α-particles; Chadwick's neutron chamber; Aston's mass spectrometer; etc. Those were the string and sealing wax days, when Nobel prizes were won with materials worth just a few pounds. We note also, when viewing the original "Wilson Cloud Chamber", by which particles flashing through space at huge speeds are detected, that C.T.R. Wilson did all the glassblowing for the apparatus himself — it brings back memories of our own youthful days of lab experiments!

Other items in the museum include a portrait of founder William Cavendish on the stairs, a bust of Maxwell, Maxwell's handwritten list of the original equipment for the laboratory, Lord Rutherford's desk, many photographs and other memorabilia.

It needs to be said that all this wealth of history is designed for physics students and professional guests — we cannot expect an active research building to have the facilities that would be needed to make it a public place. Outside visitors are therefore urged to make prior arrangement if they wish to visit the museum. Phone (01223)-337200.

Other presently active laboratories. Research in molecular biology acquired its own building and moved away from the Cavendish in 1962 — today's centre is the Medical Research Council laboratory, about two miles to the south on Hills Road, next to Addenbrooke's Hospital. The move occasioned no noticeable interruption in the work — prizes and honours have continued as before.

The University departments of biochemistry and physiology — subjects that might arguably be regarded as precursors of molecular biology — are in a large science complex bounded by Tennis Court Road and Downing Street. They may be considered as holding a slightly ambiguous position, present before the molecular biology revolution, still there now, in a sense bypassed. In fact, they too have had their own men of distinction. Gowland Hopkins (1861–1947), the first professor of biochemistry, is the man who discovered vitamins. Alan Hodgkin, a former professor of physiology, still active in research, laid the foundations of *molecular* neuroscience here by brilliant experiments (with Adrian Huxley) that define the role of Na^+ and K^+ ions in the propagation of electrical impulses along nerve fibres. Hopkins won a Nobel Prize in 1929; Hodgkin and Huxley did so in 1963. A portrait of

Hopkins hangs in a small tea room next to the biochemistry library.

3. Museums

Scott Polar Research Institute.   This institute (on Lensfield Road) is one of the most active centres in the world for polar research. A museum on the ground floor is open to the public, but is largely devoted to the heroic exploits of past polar expeditions. Scientific aspects of polar research are limited to a couple of unambitious poster displays — e.g., there was one when we were visiting on the question of whether the polar ice caps are getting thinner as a result of the greenhouse effect.

Sedgwick Museum for Geology.   This museum (on Pembroke Street) is named after Adam Sedgwick (1785–1873), who was Woodwardian Professor of Geology for more than fifty years, from 1818 until his death. Sedgwick had been a tutor at Trinity College without any training in geology when the Woodwardian chair became vacant. Bored with his job of drilling students for the mathematics tripos, he applied for the job and was accepted. Field trips at once began to take the place of dreary tutorials and an outstanding career was on its way (*see* GEOLOGY & PALAEONTOLOGY p.33).

John Woodward (1665–1728), for whom this endowed chair is named, provided the nucleus for the museum's collection of fossils, which are still to be seen in their original cabinets. Woodward had considered the fossils as testimony in support of the Biblical story of Noah's flood, a view that was of course no longer held in Sedgwick's time. The present museum has a vast display of everything from giant dinosaur skeletons to tiny molluscs. One of the best exhibits, formed by specimens collected just a few miles from Cambridge, gives a grand notion of the local fauna 120 000 years ago during the last interglacial period — hippopotamus, lion, elephant, rhinoceros; just like central Africa today!

Whipple Museum for the History of Science.   This small

but well-planned museum is on Free School Lane, close to the old Cavendish laboratory. Navigational devices, chronometers, mensuration tools, and calculating machines of a previous age are well displayed, accompanied by explanations in lay language of exactly how each device was used and its importance to people's social or economic well being. Of particular interest is the display of "Napier's Bones", a calculating device Napier invented before he thought of logarithms (*see* EDINBURGH p.282).

The Scott Museum is open Mon–Fri, 2.30–4; Phone (01223)-336540. The Sedgwick Museum is open Mon–Fri, 9–1 and 2–5, Sat 9–1 only. Phone (01223)-333456. The Whipple Museum is open Mon–Fri, 2–4. Phone (01223)-334540. Museums may be closed outside term time.

4. Other Places of Interest

The "Eagle".   Watson and Crick had lunch almost daily at the "Eagle", a pub at the north end of Free School Lane. An uncomfirmed story says that, on the day the model structure actually fell into place, the first announcement was made here, Crick telling all who would listen that they had just discovered "the secret of life". The "Eagle" has recently been enlarged, but retains some of the atmosphere of a "local". There is no picture or plaque to commemorate Watson and Crick. "Never heard of them," said the barman when one of us went in for a drink.

Francis Crick lived at 20 Portugal Place, a little to the north off Bridge Street. He renamed the house "The Golden Helix" and erected a large brass helix on the front of the house, a single helix, intended to symbolize the basic idea of helices as important macromolecular structural elements, rather than the double helix specific for DNA. Crick no longer lives here, having moved to America, but the helix remains.

COLCHESTER (Essex)

William Gilbert (1546–1603) is famous for his book *De magnete*, the first to comprehend that the earth is a magnet, that the origin of terrestial magnetism lies in its own substance and is not imposed from outside (*see* ELECTRICITY & MAGNETISM p.17). Gilbert (or Gilberd, as he signed his own name) was born in Colchester into a well-to-do middle class family. His principal profession was that of physician with a fashionable medical practice in London. Magnetism was his hobby, indulged in with his own funds and in the time he could spare from his many professional duties. He was appointed physician to Queen Elizabeth I in 1600 and continued in that post for James I, but Gilbert himself died in 1603, presumably from the plague.

Tymperleys, the fine Elizabethan timbered house where Gilbert was born, still stands on Trinity Street, added to and remodelled many times, but in its present state meticulously faithful to the original design. The property was purchased in 1956 (and saved from destruction) by a Colchester businessman, who used it to display his collection of clocks — the building is in fact now an official clock museum. (And it gives us another insight into the history of science, unrelated to Gilbert. Christiaan Huygens' invention of the pendulum clock in the Netherlands swiftly spurred the manufacture of clocks all over Europe. The Huygens design was brought to England in 1658 by Ahaseurus Fromantel, a Dutchman whose family had settled in Colchester in the previous century to escape religious discrimination — Colchester thereby became the centre for the clock industry in England.)

Gilbert died in London, but his body was brought to Colchester by his brothers for burial in Holy Trinity Church, across the street from Tymperleys. The church building is now a museum dedicated to Colchester's history — its famous oyster beds and such — but the memorial erected by the brothers over Gilbert's tomb is of course still in place. Its inscription is in Latin and unusually explicit about the interred's career — here are excerpts in English translation:

> This eldest son of Jerome Gilberd, gentleman, was born in the town of Colchester, studied the art of medicine at Cambridge, practised the same for more than 30 years at London with singular success, ... He composed a book celebrated amongst foreigners concerning the magnet, for nautical science. He

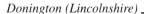

> died the year of human redemption 1603, the last
> day of November.

The memorial has a decorative border bearing the coats of
arms of William himself and numerous other members of
the family.

Trinity Museum is open Mon–Sat, 10–1 and 2–5. The Clock
Museum is open April–October only, days and hours as for
Trinity. Phone (tourist information) (01206)-712920.

DONINGTON (near Boston, Lincolnshire)

Captain Cook was not unique as a ship's captain with an
interest in science and with a naturalist on his ship's complement. Forty years after Cook and Banks we have Captain
Matthew Flinders (1771–1814) and Robert Brown (1773–
1858, *see* MONTROSE, Scotland p.295). Their voyage to the
Pacific began in 1801, their ship (*H.M.S. Investigator*) sank
en route home, but not all specimens were lost — what was
left added 4000 *new species* to the catalogue. The collection
was combined with Banks's collection in London and went
eventually to the British Museum.

Flinders was born here in Donington and a plaque marks
the site of his birthplace. A more splendid memorial is in
the parish church of St. Mary and the Holy Rood. It has a
stained glass window, a poster display and other items.
Some items are described as "on loan", but seem to have
been there for some years.

Donington is 10 miles southwest of Boston on the A52. The church
is normally open in the daytime. Phone (01775)-820418.

EAST DEREHAM (Norfolk)

William Wollaston (1766–1828), an independent and often
controversial scientific jack-of-all-trades, was born here into
a family already well known in science: his father was an

204

astronomer and a fellow of the Royal Society; two uncles were famous physicians. Wollaston's most solid work was his purification of platinum (and discovery of the related metal palladium), done in London in loose collaboration with Smithson Tennant (*see* SELBY, North Yorkshire p.247). Unlike Tennant, he worked on numerous other problems, some of them theoretical or even philosophical — he was passionately interested in Dalton's atomic theory, for example, and tried to improve on its practical applicability, probably sowing more confusion thereby than new insight. In his will he left £1000 to the Geological Society, which led to the "Wollaston Medal"; it remains an annual award of great prestige, by which Wollaston's name will always be remembered.

East Dereham is an old town, with a parish church founded in 654, where several Wollastons are recorded as former rectors. The town has some historical pride in the writer William Cowper, who came here in his declining years to wait for death. But Wollaston? All we could find was a book: *Dereham's Forgotten Scientist — William Hyde Wollaston*, by local sage Gerald Bayfield.

East Dereham is 16 miles west of Norwich on the A47.

GODMANCHESTER (near Huntingdon, Cambridgeshire)

One of the most enthralling stories of the immediate aftermath of World War II unfolded here in Godmanchester, at a large mansion called Farm Hall, owned at the time by the British Secret Service. The Germans were the first (in 1939) to demonstrate the occurrence of the nuclear fission of uranium and soon thereafter put Werner Heisenberg, one of the world's most brilliant theoretical physicists, in charge of a project to exploit the discovery. Why didn't they succeed? Why did the United States and not Nazi Germany build the first atom bomb? Was Heisenberg secretly an anti-Nazi? Did he deliberately drag his feet?

After Germany's surrender on 8 May 1945, the Secret Service whisked Heisenberg, Otto Hahn and eight other prominent physicists off to Farm Hall, detaining them there incommunicado for six months, but otherwise giving them all the freedom they could wish. Naturally, they talked mostly about physics and the war, but what they didn't know was that the building was bugged — all their conversations were recorded; transcripts of the original tapes have recently been published. What emerges is no evidence for altruism on Heisenberg's part and no mysterious reason for failure; it was incompetence more than anything else — the German calculation of critical mass, for example, was way off the mark.

The tape transcripts also reveal a certain amount of Germanic conceit. The dropping of the two atomic bombs on Japan happened in August, during the Farm Hall period of imprisonment. The Germans of course heard about it right away and their reaction was one of incredulity — how could American and British physicists have possibly succeeded where they themselves had failed?

Godmanchester is about a mile south of Huntingdon. Farm Hall (a large brick building still carrying that name) is on the left-hand side of West Street (the B1043), about a quarter of a mile from the turn-off in the town. It is now privately owned, partly as a residence and partly as business quarters.

Farm Hall. Operation Epsilon, the taping of the conversation of Germany's nuclear scientists, took place behind these walls.

GRANTHAM (Lincolnshire)

Isaac Newton went to school here from 1654 to 1660, at the King's School, a very old grammar school founded in 1328. The school still flourishes and, until recently, the "Old Schoolroom", where Newton would have actually attended classes, continued to serve its original function, but it is now used as the school library. Newton lodged with the family of apothecary Clark when attending school, next door to the George Inn on the High Street. Grantham celebrates its Newton connection with a statue (erected in 1858) on St. Peter's Hill, a small park at the centre of the town. A plaque at the King's School commemorates Newton's presence there. Other local pride in Newton is somewhat crude: an Isaac Newton shopping centre with a large and hideous plastic apple over the main entrance and "filet of pork Isaac Newton" on the dinner menu at the George, the name presumably implying apples in the recipe.

Another distinguished citizen is worth noting. Margaret Thatcher (née Roberts) was born in Grantham in the home above her father's business, "Roberts Provisions and Groceries". She studied chemistry when she went to university at Oxford and subsequently published one professional paper (on chemical reactions at a surface) before entering the political arena. The former Roberts shop now houses the waggishly named "Premier Restaurant" and there is a plaque above the entrance to note the Thatcher connection.

The Premier Restaurant is at the corner of North Parade (the B1174) and Broad Street.

HALESWORTH (Suffolk)

William Hooker (1785–1865), born in Norwich, was a precocious dilettante until Dawson Turner of Great Yarmouth took him in hand and persuaded him to settle down to serious botany. Hooker did so, buying a share in Turner's profitable Halesworth brewery to provide an income, coming to live in Halesworth (in what used to be known as the Brewery House) and eventually marrying Turner's daughter — their son Joseph was born in Halesworth in 1817. Hooker

207

became an indefatigable botaniser and writer, went to Iceland to describe its flora (the first to do so), and progressed to be an authority on mosses. He left Halesworth for Glasgow (as professor of botany) in 1820, but later returned to the south to become the driving force for the creation of Kew Gardens in London as a national establishment and a resource for botanists from all over the world. In 1841 he was appointed Kew's first official director and increased its size from 11 acres to 300 in the space of five years.

Meanwhile son Joseph was following in his father's footsteps and becoming a noted botanist in his own right: he succeeded his father as Kew director when the latter died in 1865 and further advanced its international reputation. In 1907, on the 200th anniversary of the birth of Linnaeus, the Swedish Academy awarded him a unique medal and called him "the most illustrious living exponent of botanical science". Historically, Joseph is also famous for another reason: he was a friend of Charles Darwin and (together with geologist Charles Lyell) orchestrated the joint public presentation of Darwin's and Wallace's papers on evolution at a meeting of the Linnean Society in London in 1858.

The former Brewery House still exists, now called Hooker House. It has two plaques, one each for father and son, and an unusually detailed account of their lives and careers in a poster display in the entrance hall. A tablet in the parish church of St. Mary also records their association with the town. Wedgwood medallions of the pair are kept in the vestry.

Halesworth is a pleasant small town, about halfway between Ipswich and Norwich. Hooker House presently houses offices for two dentists; it stands at the corner of Quay Street and Saxons Way, a few yards from the town's principal car park.

HEMPSTEAD (near Saffron Walden, Essex)

William Harvey revolutionised our understanding of human and animal physiology by his discovery of the circulation of the blood (*see* HUMAN BIOLOGY & MEDICINE p.7). He lies buried here in this rather remote village, about 15 miles south of Cambridge, though he probably never lived here himself. He came to be buried here because he was for the last ten years of his long life effectively a member of the families of his brothers Daniel and Eliab, successful merchants and traders, who had substantial homes in London. Eliab, perhaps imbued with a sense of history, decided to establish a country residence in Hempstead and at the same time he created the chapel in St. Andrew's Church and a vault beneath it for the burial of Harveys present and future. Sadly, the first two to be interred were two young daughters of Eliab, Sarah and Elizabeth, who died aged 13 and 9, respectively, in 1655 and 1656. William Harvey, who had no children of his own, had been very fond of them, as he was of all his nephews and nieces. When William died in 1657 (in London) the family vault was the natural place for his burial, undoubtedly agreed upon between the brothers beforehand. A large delegation from the College of Physicians came from London to attend the funeral — it was at the time an arduous two-day trip from the city.

The chapel is in the northeast corner of the church. There is a slab in the floor (with inlaid figures and coats of arms) for access to the vault — one can peer inside from a grating on the outside of the church to get a glimpse of the coffins within. A marble bust, sculpted shortly after Harvey's death, adorns the wall of the chapel and is considered an excellent likeness. The most impressive monument, however, is a splendid large marble sarcophagus — perhaps too large to be appropriate for a man who has been described as "short of stature" — which was installed in 1883 by the Royal College of Physicians and to which Harvey's remains were moved from the vault.

The coffins of Eliab and many later Harveys are in the vault, with appropriate wall memorials in the chapel above. They include another Eliab, an admiral who commanded one of the ships under Nelson at the battle of Trafalgar, and his son, Captain Edward Harvey of the Coldstream Guards, who fell (aged 22) at the battle of Burgos in 1812.

The church is normally kept open in the daytime. Phone (Radwinter) (01799)-599332.

LINCOLN (Lincolnshire)

Robert Grosseteste (*ca*1168–1253), Bishop of Lincoln, was perhaps the best of the advance guard of thinkers of his time. In the words of one historian, "he determined the main direction of physical interests of the 13th and 14th centuries." It is of course almost impossible for us to imagine ourselves as 13th century sages, before the invention of printing, when the sole source of wisdom was the church. Here is an example to give the flavour of the state of knowledge at the time: what is *vision*? Many of the ancients believed that the eye sends out rays to capture an image, but the more sophisticated notion was gaining ground that the eye is the *receiver* of signals emanating from the object perceived. Grosseteste (among others) believed this and was beginning to think about what we now call "optics", refraction of light and lenses, reflection and mirrors. The details are perhaps unimportant; what counts is that Grosseteste was a leader, a forerunner of many men of the cloth who eschewed exclusive preoccupation with salvation of the soul, extending their mission towards an understanding of the facts of nature that influence our mortal lives.

Grosseteste lived so far back in time that personal information about the man becomes scant. He was probably born in Suffolk, joined the Franciscan order, received at least part of his education in Oxford and later taught there as rector of the Franciscan school, where Roger Bacon (*see* SUNNINGWELL, Oxfordshire p.165) was one of his students. In the church hierarchy Oxford was at that time within the see of Lincoln and Grosseteste was duly promoted to Bishop of Lincoln in 1235. He died in 1253. His tomb can be seen in Lincoln cathedral, in the chapel of St. Peter and St. Paul, in the southeast transept — the chapel is also called the Students' Chapel and many academic institutions and individuals have contributed funds for its dedication and maintenance as a shrine to scholarly pursuits.

NORWICH (Norfolk)

Norwich was once the most important city in England after

London and can boast distinguished scientists in its history, though they are scarcely remembered today: a book for tourists (Historic Norwich) pictures 8 "famous sons and daughters", but no one mentioned here is included.

Church of St. Peter Mancroft. Thomas Browne (1605–1682) was an early natural philosopher, trained as a physician and pharmacist. He was only peripheral to science per se, in no way an innovative experimentalist, but he contributed much philosophically as one of the early English polymaths who broke loose from the medieval form of scholarship and forcibly advocated a more modern approach. All things were to be questioned; observation and experiment were needed to settle questions about nature; reasoning and logic should be applied to all problems. His conclusions were as often wrong as right, but his skill as a writer and communicator was instrumental in influencing others away from the acceptance of revealed truths towards objective investigation of the natural world. He settled in Norwich in 1636 (at age 31) and lived here for the rest of his life. He is buried in the chancel of this great church, with a memorial inscription on the wall above and a statue on the lawn outside.

Castle Museum. Samuel Woodward (1790–1838), geologist and antiquary, was almost wholly self-educated. His father was a weaver and Samuel was sent to work in the same industry at the age of seven. His interest in natural history and archaeology was aroused by the reading he did in his spare time, encouraged by a local alderman. He had an extensive collection of fossils and antiquities from the Norfolk region and published a number of articles in respected journals. His three sons continued his interest in local geology, and the youngest, Henry Woodward, had a distinguished career and became president of the Geological Society of London, winner of the Wollaston Medal, and editor of the *Geological Magazine.* Even a grandson followed him into the field of geology. Woodward's extensive collection was purchased after his death by the Norwich museum and whatever portion of it has survived is now in the Castle Museum, which has a reasonably good modern geology exhibit. The only mention of Woodward, however, is on a poster in one of the display cases.

Other scientists, scarcely recognised.   The most influential
of the Norwich scientists was James Edward Smith (1759–
1828), the naturalist who brought the Linnaeus collection
from Sweden to London (saving it from decay in the hands
of indifferent heirs) and founded the Linnean Society of
London for promotion and extension of the Linnean system.
Smith was born in Norwich, in a house on Gentleman's
Walk, where his father had a draper's shop. He horrified his
London associates when he decided in 1797 to move back to
his native city — he would be "buried alive", in the words
of one of his friends, in what had become by then a decid-
edly provincial city. Smith lived in Norwich nine months out
of the year thereafter, in a substantial terrace house on 29
Surrey Street; the Linnean collection went with him and
remained in the house till it was transferred back to London
after his death. The house bears a plaque in Smith's memory.

One other name worthy of mention is that of the botanist
William Hooker (*see* HALESWORTH, above), who was born in
Norwich in 1785 and began his education at the local grammar school — surely meriting inclusion as a "famous son".
We actually found a tiny remembrance. We were in the
keeper's office at the Castle Museum (talking about
Woodward) when we spotted two plaster busts on a window
sill, behind a stack of discarded boxes — they turned out to
be Hooker and Smith, rescued from a storage room that was
being converted into another office.

Gentleman's Way is on the Market Place opposite St. Peter
Mancroft Church, Surrey Street is about 400 yards to the south
and the castle is even closer, on high ground to the east. The
Castle Museum is open daily all year; Phone (01603)-223624.

REVESBY (near Horncastle, Lincolnshire)

Joseph Banks (1745–1820) of Revesby is best known as the
leader of the party of scientists and assistants who accompanied James Cook on his first voyage of discovery.

Note the name Botany Bay for the expedition's first anchorage in Australia — symbolising the party's perceived mission. Banks later became notable on many other counts, as a major influence behind the creation of Kew Gardens in London, for example, and as the longest-serving President of the Royal Society (42 years from 1778 to 1820). His specimen collection was the nucleus which ultimately grew into the Natural History branch of the British Museum: the museum library is the repository of Banks's correspondence.

Though Banks's principal residence during his active career was in London, Revesby Abbey was the true family home — a grand residence by all accounts, whose name derives from its location, on land that formerly was part of the estate of a Cistercian abbey. Banks returned here every summer, taking part in the village fair and in other ways asserting his role as hereditary local squire. When he was raised to the peerage in 1781 he became the Baronet of Revesby Abbey — the first and last of the line, for he had no direct heirs (_see_ PULBOROUGH, West Sussex p.114). Banks's mansion was replaced by a yet fancier building in 1849, and even that is now delapidated and no longer habitable — but the exterior is still in good shape and reminds us of the former splendour. Next to the church on the village green is a row of almshouses, built by Banks's father in 1728 and authentically restored in 1987.

Revesby is 7 miles south of Horncastle, at the intersection of the B1183 and A155. A fine view of the mansion is obtained through gates on the A155, less than half a mile east of the village. The village fair goes on, held annually in early August.

TERLING (near Chelmsford, Essex)

Terling Place, the impressive country mansion of the Rayleigh family, is where the physicist, the third Lord Rayleigh (1842–1919), lived and worked and was buried after he died. Although Rayleigh served at different times as Cavendish professor at Cambridge and professor at the Royal Institution in London, Terling was his scientific base. He converted one wing of the grand mansion into a laboratory, filled with equipment purchased with his own money

— he even had an expensive telescope installed, projecting upward through the baronial roof. He was a polymath and his work impinged on several branches of physics and even chemistry — from theoretical work on sound and other wave motions to the discovery of argon (for which he earned a Nobel Prize) and to the measurement of actual molecular dimensions. He was also a rare model of altruism, putting in long hours of service to the Royal Society and in helping individuals who needed a boost to get their work published.

Rayleigh was buried at Terling, in the southeast corner of the churchyard, and the funeral was attended by all the principal officers of Cambridge University and of the Royal Society. Two years later a memorial tablet with a medallion portrait was put up in Westminster Abbey; there is also a mural tablet in Terling church.

Terling Place, still residence of the current baron, also houses "Lord Rayleigh's Farms", the major source of the third lord's income and to this day a bustling and presumably profitable business enterprise. They used to sell dairy products directly to the public through a chain of retail outlets in London — these were sold in 1928, but the farm's lorries, delivering produce wholesale to food shops, can still be seen on London streets.

Terling is best approached on an unnumbered road from Hatfield Peverel, 5 miles east of Chelmsford. Only the church and churchyard are open to the public.

WALTHAM ABBEY (Essex)

The line of the Greenwich meridian is marked in many places from Peacehaven (East Sussex) on the English Channel to beyond the Humber estuary in the north. Of all these places, the most enjoyable to visit is Waltham Abbey, where the historic and beautiful abbey church lies adjacent to Lee Valley Park, a 23-mile stretch of countryside and nature preserve along the Lea River — no one knows what bureaucratic quirk led to the difference in spelling between the river and the park. The meridian is marked by a mosaic on the paving of the walkway from the principal car park to the information centre and shop. It continues to an ancient stone bridge across Cornmill Stream and coincides beyond (via an underpass under a highway) with a long footpath, officially designated as "Meridian Walk".

Of special interest, just beyond the east end of the present abbey, is the probable position of the grave of King Harold, who was left dead on the battlefield of Hastings in 1066, but brought here by his wife, to lie in a place he loved and where he had found religious solace. The grave is not actually on the meridian, but close (about 100 yards away) — would there have been some symbolic significance if Harold's remains were lying today with the skull in the eastern hemisphere and the feet in the western?

Waltham Abbey is on the A121, just off the M25 motorway. Follow signs to Lee Valley Park.

WOOLSTHORPE (near Grantham, Lincolnshire)

Isaac Newton was born on Christmas Day in 1642, at Woolsthorpe Manor, now preserved as a monument by the National Trust. (He was educated at King's School in nearby Grantham.) The manor is not in any sense a museum, but simply a house — a farmhouse really, for "manor" is too grand a name — that has been preserved. It does contain a replica of Newton's device for producing the coloured spectrum from white light, an electric "candle" replacing the sunlight or wax candle light that he himself used, and a descendant of the famous apple tree stands on the lawn outside. The house has some interest for its own sake — there is an "owl hole" under the eaves, for example, to provide a nesting place for owls, with the intent that they should keep the place clear of mice and other vermin.

One must remember when one visits that this is not merely Newton's place of birth, for he returned to Woolsthorpe to live with his widowed mother after his graduation from Cambridge. It was the time of the Great Plague throughout Europe (31 000 people perished in London alone over a two-year period) and Cambridge was badly affected, not a healthy place to be. Newton thus spent the years 1665–1667 in virtual seclusion at Woolsthorpe,

Newton's birthplace, Woolsthorpe Manor, from a painting by J.C. Barrow in 1799. The house is now a National Trust property.

during which time he formulated the binomial theorem of mathematics, developed calculus, studied decomposition of white light into its spectral colours by means of a prism, and began his study of mechanics, including the universal law of gravitation. Surely two of the most productive years ever spent by a single scientist, notwithstanding the fact that none of the work was published until later, some of it *much* later — Newton's fear of adverse criticism is well known.

Newtons's only near-contemporary rival in the field of physics (if, that is, he had one at all) was the Dutchman Christiaan Huygens and a fascinating contrast emerges if one can combine a visit to Woolsthorpe with a visit to the Huygens family summer home Hofwijck on the outskirts of Den Haag — or, at least, if one can remember one when one visits the other. For Christiaan Huygens had a truly aristo-cratic family background, with a grandfather who had been secretary to the Dutch national hero, William I of Orange. And the old summer home (now the official Huygens muse-um) is an appropriately elegant house, set among orchards and decorative canals, with fine furnishings, clearly designed for elegant garden parties and musical soirées, attended by well-dressed gentlemen and ladies with long dresses and delicate slippers. The Newtons were not poor — they were country squires and their family home in

Woolsthorpe was substantial. But their house was clearly a farmhouse, designed for people who work all day with muddy boots.

Woolsthorpe is 1 mile off the A1: follow signs to Colsterworth. The house is open to the public from April to October, Wed–Sun and Bank Holidays, 1–5. Phone: (01476)-860338.

7
Northern England

BURNLEY (Lancashire)

Towneley Hall and the Towneley family — Catholics, who refused to renounce their allegiance when Protestantism became the official religion — have been an integral part of Burnley history for 500 years. One of them, Richard Towneley (1629–1707) was a distinguished member of the scientific community of his day. His father Charles had fought to preserve the rule of the Stuart kings and was killed at the battle of Marston Moor, the same battle in which William Gascoigne was killed (*see* MARSTON MOOR, p.240). Gascoigne's papers and his prototype micrometer for the telescope passed into Richard's hands and it was at Towneley Hall that the instrument was actually perfected and first applied to astronomical measurements — after 1670 Towneley collaborated with the Astronomer Royal (John Flamsteed) on routine astronomical observations.

Richard Towneley, denied participation in government on account of his religion, devoted most of his energies to science, much of it outside the field of astronomy. He is noted for his barometric measurements and meteorological observations, for example, and in that connection he and a fellow philosopher (Henry Power) recognised the inverse dependence of the volume of a fixed quantity of air on its pressure — essentially what we know as Boyle's Law. He corresponded with Boyle about this and his contribution is acknowledged in Boyle's publication of the relation.

Towneley Hall remained in the hands of the family until 1901, when it was sold for a nominal sum to the Burnley Corporation. It is now a museum and the grounds are a huge recreation area which includes two golf courses and other facilities. The Hall has of course been altered over the years, but the basic structure is unchanged and some of the rooms are probably not very different now from the way they were back in the 1600s. The wing of the building used for Richard's experiments, however, was demolished in one of the reconstruction projects and the museum has no exhibits related to the scientific work that was done there. Nevertheless, we found that Towneley Hall gives one a unique feel for the opulent surroundings in which science was often carried out "way back when".

Townley Hall is on the Todmorden Road (A646) and is open all year Mon–Fri, 10–5; Sun 12–5. Phone: (01282)-424213.

BYERS GREEN (near Bishop Auckland, Durham)

The Milky Way is our galaxy — a commonplace notion today, but pure fantasy when Thomas Wright (1711–1786) of Byers Green proposed it in 1750, in his book, *An Original Theory or New Hypothesis of the Universe*. Wright was a singular and sometimes confused person, with a speech impediment that interfered with his getting more than a rudimentary education. He became immersed in the world of astronomy while still a teenager. He had an obsession, to relate astronomical facts to God and the creation, which led him to the imaginative idea that God's place, the moral centre of the universe, should also be the gravitational centre, the centre of physical order. Since the sun was much too hot to be the "place" of any reasonable God, this meant that the sun and the stars (to prevent gravitational collapse) had to be in continuous quasi-planetary motion about some other "divine" spot in the universe. Looking at the sky for inspiration, he imagined the Milky Way as visual evidence for a swirling mass of stars — it took him some years to realise that it must then be a thin *disk-like* assembly, but eventually he did and published his book. Among those who read it was the German philosopher Immanuel Kant, who incorporated a disk-shaped model of our galaxy into his own cosmological ideas and thereby gave Wright a wider audience. Wright had other ideas about stars which were more fanciful and did not stand the test of time; he can never be cited for rigorous or critical analysis. But we must remember that he was more than 50 years ahead of his time in giving any rational thought at all to the cosmological significance of the Milky Way.

Wright was born in Byers Green and spent most of his years there. He built himself a fine villa and created an observatory on high ground on the village green in nearby Westerton, which was not actually completed till after his death. The observatory tower still stands, known locally as the "Westerton Folly" and it is marked by a stone tablet. The villa is gone, but its former location is marked by a plaque.

Byers Green and Westerton are about 4 miles northeast of Bishop Auckland, on minor roads off opposite sides of the A688. The plaque in Byers Green (put up at the urging of the former

"Westerton Folly", the observatory tower built by Thomas Wright.

Astronomer Royal, Arnold Wolfendale) is inside the front garden of a bungalow ("Hallgarth") at the foot of Chent Street — the debris from the bulldozed villa, said to have included an elaborate astronomical scheme, lies buried in a pit behind the bungalow.

DENT (near Sedbergh, Cumbria)

This village with cobbled streets, at the edge of the Yorkshire Dales National Park and accessible only via a narrow curving road from Sedbergh, is the birthplace of geologist Adam Sedgwick (1785–1873, *see* GEOLOGY & PALAEONTOLOGY p.33; also *see* CAMBRIDGE p.201), son of the local vicar. It is a spectacular place, reminiscent of a French or Swiss mountain village, which must be seen to be fully appreciated, not only for its scenery but also for its remarkable (and not boastful) display of "Sedgwickism". There is a memorial fountain for Adam in the centre of the village, backing onto a simple rough-hewn block of granite. The Norman village church (founded around 1080) also has a memorial to him, a stained glass window and a marble tablet explaining that he was baptised here but buried in Trinity Chapel in Cambridge. But this is only one of a profusion of monuments to various Sedgwicks in the church, some of them vicars, others just benefactors to church or community — one of them, a later Adam Sedgwick, was professor of zoology at Imperial College in London. The Sedgwicks remain prominent in the village to this day: the local plumber, for example, has a van boldly marked with the family name. The lady at the post office of whom we made some inquiries admitted that she, too, had been born a Sedgwick.

Adam himself, it might be noted, never abandoned his local connections after he became famous. He often returned to the area and was president of the Kendal Literary and Scientific Society for 32 years.

Sedgwick Geological Trail. This nature trail, created and maintained by the Yorkshire Dales National Park, is designed to explain the geology of the "Dent fault", which was first discovered and interpreted by Sedgwick. It may

A block of granite commemorates Adam Sedgwick in the picturesque village of Dent.

require some previous knowledge to be appreciated — our untrained eyes found it a little hard to match some of the features described in the trail guide pamphlet with what we could actually perceive.

Dent is 5 miles south of Sedbergh on an unnumbered road. The geological trail is below a viewpoint and car park at Longstone Common, about a mile east of Sedbergh on the A684.

EAGLESFIELD (near Cockermouth, Cumbria)

One never ceases to be amazed at the lack of intellectual backwaters in the England of earlier centuries. Scholars and literary figures turn up no matter how far we stray from the main road. The village of Eaglesfield, in an idyllic setting at the edge of the Lake District, is a good example, for it is the birthplace of John Dalton, one of the great names in the history of chemistry (*see* CHEMISTRY p.43). The former Dalton house, a whitewashed bungalow, still stands and bears the commemorative inscription "John Dalton DCL.LLD — The Discoverer of the Atomic Theory — was born here Sept. 5 1766, died at Manchester July 27 1844". The house is a private residence and not open to the public.

The poet William Wordsworth was born in 1770 in Cockermouth, just a couple of miles away. His father was business manager for Sir James Lowther and the Wordsworth home, rented from Sir James, is a more substantial house than Dalton's. It is now a National Trust property, retains the original staircase and fireplaces, and is open for public view. Both Dalton and Wordsworth retained a life-long love for the Lake District. Dalton returned to the area each year for his summer holiday and Wordsworth came to live there permanently (at Grasmere, further south) and even wrote a travel guide to the Lakes.

Pardshaw Hall. Dalton died and was buried in Manchester, but there is a memorial to him in a most unusual graveyard at Pardshaw Hall, about a couple of miles from Eaglesfield. The graveyard is part of an ancient meeting house of the Society of Friends (also called *Quakers*), a Christian sect of which Dalton had been a member. The small cemetery has been designated as a national cemetery — a Quaker who dies in a place that lacks a cemetery may be brought here for burial. The Society erected the memorial to Dalton 50 years ago. "He was not for an age, but for all time" is what the simple inscription tells us.

Eaglesfield and Pardshaw Hall are reached by taking the A5086 south from Cockermouth. Eaglesfield is on a turning to the west, Pardshaw Hall (distinct from the village of Pardshaw) is the next turning on the opposite side. It is just a litle row of houses and the meeting house is the first building on the left as one enters. It is used for services only once a month and the graveyard is hidden from view by a high wall. Phone (to get keys): (01900)-826365.

Kendal (Cumbria)

Kendal was virtually a Quaker town in the 18th century and beyond and a legendary figure from that period is John Gough (1757–1825), the blind philosopher. He had small-

The house where John Dalton was born. The plaque is above the doorway on the right of the house.

JOHN DALTON. D C L. LL D.
THE DISCOVERER OF
THE ATOMIC THEORY
WAS BORN HERE SEPT 5 1766
DIED AT MANCHESTER JULY 27 1844

pox when less than 3 years old, which destroyed his sight, but he nevertheless became a huge influence in this area, most emphatically so in relation to John Dalton, who was his pupil for 4 or 5 years and for whom he got the job that enabled Dalton to move to Manchester (*see* CHEMISTRY p.43). A few years later Gough also taught William

Whewell, subsequently the much celebrated master of Trinity College, Cambridge.

Gough did experiments, too, despite being blind, and made a discovery that would 50 years later be judged to be quite remarkable. He found that a rubber band (held between his lips) becomes warm when stretched and cools again when relaxed — the reverse of what happens with most materials. Joule and Kelvin confirmed the observation in 1857, by which time a corollary conclusion could immediately be drawn (and confirmed), that rubber, in contrast to most materials, has a *negative* coefficient of thermal expansion — it contracts when heated. Another hundred years later, when *polymer science* became a theoretically and practically important field, this property of rubber became a key element in the molecular interpretation of the behaviour of polymer fibres. The "bible" of polymer science, *Principles of Polymer Chemistry*, written by Paul Flory in 1953, devotes several pages to Gough's experiments, even quoting directly from the article published by him in the Proceedings of the Literary and Philosophical Society of Manchester.

The Kendal Museum is fully cognisant of the town's importance in the history of science and prominently displays busts and explanatory posters, not only for John Gough and John Dalton, but also for Thomas Gough (John's son), who was a lifelong friend and collaborator of the geologist Adam Sedgwick, and for Sedgwick himself, who was born not far from Kendal (*see* DENT, above p.222). The meeting house, where Quakers have worshipped since 1688, still exists and is still used for religious observance. The Quaker school, where the Goughs and Dalton taught and learned, was in a house across the street from the meeting house, now no longer used as a school. The meeting house has an unusual modern work of art, a tapestry that displays the history of Quakers in Britain. It includes a panel devoted to Quaker scientists — Dalton among them, of course, and also the cosmologist Arthur Eddington (*see* ASTRONOMY p.30), a native of Kendal, who, however, moved away to Devon at an early age.

The Quaker meeting house is on Stramongate. The museum is on Station Road and is open Mon–Fri 10.30–5 and Sat and Sun

2–5. (Sat 10.30–5 in summer). Phone (museum) (01539)-721374.

KESWICK (Cumbria)

Another example of intellectual ferment in the Lake District — where one of the lakes (Derwentwater) plays a direct role! It was 1772 and the future local sages, John Dalton and William Wordsworth, were still children. Benjamin Franklin (then resident in London) was on his way to Edinburgh and en route he made two recorded visits, one to John Smeaton near Leeds, the other to William Brownrigg of Ormathwaite, a hillside estate above Keswick. In both places he was eager to demonstrate an experiment he had done on a pond at Clapham Common in London — the spreading of oil on water and the concomitant stilling of the waves that had ruffled the pond's surface. Here the demonstration was done on Derwentwater from a boat.

The spreading of the oil is spectacular — just a teaspoonful expands to cover half an acre — and the result here was no different from what it had been in London. What was exceptional was the scepticism of a vicar in Carlisle (James Farish), who thought the report he had heard was surely exaggerated and challenged Brownrigg on the subject. Brownrigg, himself an amateur experimenter and Fellow of the Royal Society, thereupon urged Franklin to publish a full account, which he did, satisfying Farish, but in the process also leaving a permanent record in the scientific archives, which we would not otherwise have had. The experiment (repeated many times since on a lesser scale) is at the heart of the subject of surface chemistry — the spreading results from the fact that the oil on the surface forms a layer that is just one molecule thick. (And the thickness of the layer — less than one ten millionth of an inch — provides an actual measurement of a molecular dimension, though that was not appreciated for another 100 years!)

Ormathwaite is still there, surrounded by a stone wall, on an unnumbered (signposted) road off the main road to Carlisle (A591). The view of Derwentwater from here is said to be the best obtainable.

KIRBYMOORSIDE (North Yorkshire)

Kirkdale cave, where William Buckland in 1821 identified the bones of all sorts of exotic species — lions, elephants, hyenas, etc. — is on a by-road about a mile west of Kirbymoorside, at the point where the road fords Hodge Beck, a normally dry stream with its main course underground. The cave opening is now well above ground in a limestone cliff and one can look through the opening to see the spacious interior, going back 20 or 30 yards into the rock. The entrance is only three feet high, which seems a little low for elephants and led Buckland to conclude that the cave must once have been a den for hyenas, who dragged in remains of larger animals piecemeal for food. Buckland accepted his finds as prima-facie evidence for the Biblical narrative of the Flood; they were a major stimulus for his best-selling *Reliquiae Diluvianae*, published in 1823.

The cave is on the east side of the ford and one should use care in climbing to the opening because the rock is crumbly. On the opposite bank of Hodge Beck is St. Gregory's Minster, a well maintained Saxon parish church with a perfectly preserved sundial above its doorway, dating from before the Norman conquest.

LANCASTER (Lancashire)

One of Lancashire's best known scientists was the chemist Edward Frankland (1825–1899), born in Catterall about midway between Lancaster and Preston. He was a man who flitted from pillar to post, not only in terms of the positions he held, but also in terms of the subject matter of his scientific work — a man whose historical importance rests more on his *influences* than on solid, completed projects. He is cited most often for work he did in Manchester in 1852 on "radicals" in organic chemistry, which led him to novel ideas about the combining powers of atoms. He was the first to have an inkling of what is now called atomic valence — each atom has "only room, so to speak, for the attachment of a fixed and definite number of atoms of other elements", to quote his own words. But the proper atomic weight for carbon was not yet established at the time, so that quantitation was all but impossible, and he himself did not subsequently make serious specific contributions to the subject.

Frankland worked for the government as an advisor on water analysis and river pollution — concern about this is not as recent we may think. He was an active teacher, sometimes holding two teaching appointments simultaneously, and had many friends. One of them was John Tyndall, with whom he was co-teacher at Queenwood College (*see* EAST TYTHERLEY, Hampshire p.106). He and Tyndall were avid mountain climbers and even did an experiment together in 1859 on the summit of Mont Blanc to study the effect of atmospheric pressure on combustion.

Other noteworthy Lancastrian scientists — actually born in the city — were the Darwin antagonist, Richard Owen (*see* EVOLUTION & ANTHROPOLOGY p.38), and Cambridge master William Whewell, who coined many now familiar words, e.g. scientist, physicist. Together with Frankland, they are remembered on an ornate memorial for Queen Victoria on Dalton Square, but only in company with many other famous Victorians. No more explicit manifestation of public pride is in evidence. The city museum once displayed some of Frankland's chemical apparatus, but no longer does so.

LEEDS (West Yorkshire)

City Square. Joseph Priestley (1733–1804) lived here, twixt church and brewery, and gained wisdom from both! Priestley was born in Birstal Fieldhead, just outside Leeds, and returned to the city in 1767 as minister of Mill Hill Chapel, which stood here on the same site as the present (rebuilt) Unitarian chapel. It has a "Priestley Hall" in the back for church functions. Next door is a throroughly modern office building, Priestley House, which is said to be on the site of Priestley's home when he was minister. It carries an explanatory plaque.

What the plaque neglects to tell us that there used to be a brewery on the other side of Priestley's house and that Priestley found it just as interesting as his church, not necessarily for its liquid product, but certainly for the gas above the fermentation vats, which was quite different from ordinary air, unable to maintain the burning of a candle, but clearly the same gas that was present in the mineral waters that people drank for the sake of real or imagined benefits to their health. Priestley got the initial ideas here that led to his later invention of artificially carbonated water — which earned him the Royal Society's Copley Medal — and to his subsequent research on gases in general, highlighted by his discovery of oxygen (*see* CHEMISTRY p.42).

City Square itself is a likeable place, designed to be both commemorative and decorative. It has statues of Priestley, James Watt and John Harrison and (in the centre) Edward, the Black Prince, the "flower of English chivalry". To balance all this weighty masculinity we have eight scantily clad nymphs, four labelled "Morn" and four labelled "Even".

Austhorpe. John Smeaton (1724–1792), who lived all his life in the Leeds suburb of Austhorpe, illustrates the need for a close relation between basic science and industrial projects, which we nowadays take for granted — with this difference, that in the 18th century the same person had to do both, there being no reservoir of professionals in applied science to whom an aspiring entrepreneur could turn. Smeaton was a respected fellow of the Royal Society (winner in 1759 of its Copley Medal) and at the same time an engineer who engaged for profit in huge engineering projects, such as building bridges and power stations. Smeaton in fact coined the term "civil engineer" to distinguish himself and his kind from army-trained military colleagues. To illustrate Smeaton's closeness to very basic science, we should point out that he was the first person to create the concept of engine efficiency — how much work you get per unit of fuel — and to make calculations of it. He called it "duty", a term still used by engineers today.

Smeaton's most celebrated exploit was the construction of the vital Eddystone Lighthouse, 14 miles out to sea, south of Plymouth. Smeaton carried out an exemplary study of the effect of the formulation of cement on its ability to set and survive under water, varying one component at a time while others were held constant, before embarking on the project. Previous lighthouses at the same site were short-lived;

Smeaton's survived for 120 years and was removed in 1882 — because the rock on which it stood had eroded! The lighthouse itself was still strong.

There is a fine memorial to Smeaton in the chancel of the Parish Church of St. Mary in Whitkirk (an area adjacent to Austhorpe), where he is buried. It exults in his success with far away Eddystone light, "where one had been washed away by the violence of a storm and another had been consumed by the rage of fire".

The Church of St. Mary is on the major road, the A63, on the Leeds side of Austhorpe, at a traffic light marking the turn-off towards the park and art museum at Temple Newsam.

LIVERPOOL (Merseyside)

Merseyside, from Liverpool and Birkenhead to Widnes and Runcorn, can legitimately claim that it laid the foundations of Britain's chemical industry. The industry remains prominent today. The Unilever complex, across the river from Liverpool at Port Sunlight, is the world's leading manufacturer of soap and detergents — and famous as well for the model community for the company's workers, set up a century ago by Lord Leverhulme himself. The place to go to absorb the atmosphere and historical background and technical information, too, is the Catalyst Museum in Widnes, which (like Liverpool itself) was formerly part of Lancashire, but it is now declared to be neither in Lancashire nor in Merseyside, but in Cheshire — see under MIDLANDS (NORTH) for a visit.

MANCHESTER (Greater Manchester)

The city of Manchester is a product of the industrial revolution and until quite recently its image has been one of grim pollution, coupled with gross social inequality — wealthy barons of industry on the one hand, masses of underpaid workers in unsanitary housing on the other. As late as 1890, the life expectancy in Manchester was six years below the average for England and Wales.

But even at the height of the urban squalor there was always another side to Manchester, which has received less publicity. For two centuries it has nurtured a tradition for scientific research and discovery — not, as might have been expected, *industrial* science tied to its smoky factories, but *pure* and *fundamental* science. The city has been especially outstanding in the area of atoms and energy, the part of physics that impinges most closely on chemistry, where it can claim three of the all-time greats: John Dalton (1766–1844), James Prescott Joule (1818–1889) and Ernest Rutherford (1871–1937). Dalton created the working model of the atom here, pretty much the way that laboratory chemists think of it to this day, little spherical bodies, combining with each other in specific integral ratios to form "molecules" (*see* CHEMISTRY p.43). Joule did more than anyone else to establish the concept of a universal thing called "energy", demolishing decades of misconceptions about the separate identities of heat, electrical energy, etc. (*see* HEAT & THERMODYNAMICS p.20). Rutherford is the father of the "modern" atom, explaining in a sense how Dalton's atoms work when they combine, but also going beyond that in providing one of the opening chapters for our current world of subatomic particles and megabuck accelerators to study them. (*See* INSIDE THE ATOM p.23.) No other provincial city in Europe can rival Manchester in this record of providing so many of the basic milestones for our understanding of the material world around us.

Entries for Manchester are arranged as follows:

1. City Centre
2. Museum of Science and Industry
3. University of Manchester
4. Salford and Barton

---

1. City Centre

Town Hall.   In the centre of Manchester stands the Town Hall, a massive Victorian Gothic structure with an extravagantly spacious interior and a proud clock and bell tower on the outside. It was built in 1877 and it proclaims in no un-certain fashion that "Manchester has arrived", with remark-able emphasis on science as integral to the city's history. We see it right from the start, as we enter the building from Albert Square, with statues of John Dalton and James Prescott Joule just inside the door, one on the left and one on the right. Even more impressive is the Great Hall upstairs, which is adorned with twelve mural paintings by the celebrated pre-Raphaelite painter Ford Madox Brown. They took fifteen years to complete and give a panorama of the city's history — the building of a Roman fort at "Mancenion", the expulsion of the Danes in the 10th cen-tury, the foundation of the textile industry by Flemish weavers, and so forth. Two of the twelve panels deal with pure science. One shows John Dalton collecting marsh-fire gas, watched by a group of children and a cow peering curiously across a fence. The second refers to an even earlier time (1639, *see* MUCH HOOLE, p.242) and pictures William Crabtree observing the transit of Venus across the sun — not quite accurately, for Crabtree was a young man in his 20s at the time, whereas the mural makes him look like an old sorcerer. Another mural represents a major tech-nological achievement, the opening of the Bridgewater Canal in 1761.

The Great Hall in the Town Hall is used for banquets and other functions and may be closed to the public at unpredictable times. Phone (0161)-2363377, extension 309/310.

Portico Library and the "Lit and Phil".   The organisation which has played so great a role in the cultural history of Manchester — the intellectual home of both Dalton and Joule — is the Manchester Literary and Philosophical Society, popularly known as the "Lit and Phil". Founded in 1781, it still flourishes, still publishes its *Memoirs*, still holds weekly meetings to widen "public appreciation of any form of literature, science, the arts and public affairs" (party politics explicitly excluded). It is completely independent, relying entirely on subscriptions and other private support, and is run from a small quite ordinary office. Many of its meetings have for decades been regularly held in the Portico

John Dalton collecting marsh-fire gas. A mural painting by Ford Maddox Brown in the Manchester Town Hall.

Library, another independent relic of the time of Dalton and Joule, both of whom were members. Its single large room, surmounted by a Georgian dome, is an upper floor of the original building; the bookshelves used to be a gallery around a larger reading room on the ground floor. Note the inscriptions above the shelves, e.g. "Polite Literature", "Voyages and Travels". It is a great place to imbibe the Victorian atmosphere, like something out of a Dickens novel, but in three dimensions.

It is worth noting that the first secretary of the Library (and, of course, a prominent member of the Lit and Phil) was Peter Mark Roget (1779–1869), a Manchester physician. He lived to very old age and is best known to the English-speaking world for his *Thesaurus*, which was a "retirement project", compiled after he quit his medical/academic career.

The Portico Library is at the corner of Mosley and Charlotte Streets and is open Monday to Friday. Phone (0161)-2366785.

St. Ann's Church: Joule and the Conservation of Energy. James Prescott Joule was still alive when Ford Madox Brown was doing his work, so that Joule's contribution to Mancunian science was too recent to be recognised in the retrospective view of the paintings. His first public presentation of his epic experiments on heat and work in 1847 was actually given in the reading room of St. Ann's Church, which stands on the square of the same name. Historians of

Statue of James Joule at the entrance to Manchester Town Hall.

science often refer to the lecture as the "St. Ann's Church Lecture".

The lecture on the same subject to a national audience came a month later at a meeting of the British Association in Oxford. It was scheduled for late in the day and the chairman asked Joule to keep it brief, which he did. One young man in the audience (William Thomson, later Lord Kelvin, but at the time an unknown) appreciated the importance of the paper and asked leading questions from the back of the room — without him the dissemination of Joule's results might have taken much longer than it did.

Dalton Homes and Funeral. Dalton made his living largely by private tutoring. He was a bachelor and lived frugally, always within a few steps of the Portico Library. For 26 years he roomed with the family of the Revd William Johns at 10 George Street; he lived alone for the last years of his life, at 27 Faulkner Street — now part of Manchester's picturesque Chinatown. Dalton tutored and did many of his experiments in a building belonging to the Society of Friends at 36 George Street; the site is marked by a commemorative plaque.

A fine statue of Dalton stands outside the Dalton building of Manchester Metropolitan University at the corner of Oxford Street and Chester Street, a little south of the city centre. Set in the ground in front of the statue are two large stones that covered Dalton's grave at the former Ardwick Cemetery, now displaced by the needs of urban development. Dalton had been given a public funeral after his death in 1844. Thousands of mourners filed past the body as it lay in state in the Town Hall and joined the funeral procession to Ardwick.

2. Museum of Science and Industry

This museum is based in the buildings of a former railway station. Despite its name, it bears little relation to science, concentrating instead on Manchester's industrial history: machine tools, Rolls-Royce cars, etc. "Thrill to the sounds of industry" and "Wonder at the miracle of flight" were typical promotional themes when we visited a few years ago. Dalton, Joule and Rutherford are mentioned in the introduction to the museum's souvenir guide, but there are no matching exhibits within. On a more recent visit we saw a modest but accurate poster display, entitled "Two centuries of Manchester science" and attributed to the museum — but we found it not here at the museum, but on loan to the Chemistry department of the University (*see* below).

We do recommend a visit to the old terminus of the Bridgewater Canal, just opposite the museum on Liverpool Street. This was the first canal in England to be constructed (1759–1761) across open country independent of any river bed; its purpose was to bring coal from the Duke of Bridgewater's mines into the heart of the city, where it went to feed the hungry furnaces of industrial production. It has been described (by D.S.L. Cardwell) as "the first major event in what we call the Industrial Revolution"; no one can question its importance as a stimulant to innovative thinking of all kinds and its role in Manchester's uniquely eager acceptance of progressive scientific ideas. (Note that it preceded the founding of the Lit and Phil in 1781.)

The museum is on Liverpool Street, at the southwest edge of the city centre, in the Castleford area, which is being developed as a kind of heritage centre. It is open every day all year (except Christmas) 10–5.

3. University of Manchester

The University grew out of Owens College, founded in 1851 as the result of a large bequest in the will of John Owens, a wealthy local business man who was unmarried and had no direct heirs. It almost failed after a few years and was rescued by the advent of Henry Roscoe as professor of chemistry. He saved the college and brought in more students and financial support by unprecedented emphasis on chemistry and other sciences over all other subjects — science has been the university's pride ever since; we can list here only a fraction of the oustanding people who have passed through its laboratories and classrooms.

Rutherford splits the atom. Arthur Schuster (1851–1934) first came to Owens College as a demonstrator for one term in 1873 and returned in 1881 as professor. His own research was in the field of spectroscopy, carrying to Britain from Germany the then new discovery that the atoms of each distinct element were associated with spectral emission lines at specific wave-lengths. Perhaps more important, he was the force behind the creation of the first experimental physics laboratory and he brought Ernest Rutherford to Manchester from Canada. Rutherford was professor here from 1907 to 1919 — this is where the modern picture of the inside of an atom first emerged and where the seeds of purposeful atomic disintegration were planted. Rutherford attracted a host of brilliant coworkers, not the least of whom was Niels Bohr from Denmark, who (among other achievements) provided the theoretical link between Schuster's atomic spectra and subatomic structure.

Manchester's successive physics buildings have been named after Schuster. The original building on Coupland Street, where Rutherford worked, is now used by the Psychology and Sociology departments, but Rutherford's laboratory still exists in the basement — boarded up, no sign to indicate its former illustrious inhabitants, but the porter will show you where it is if you ask him. There is a plaque on the main road nearby, at the corner of Coupland Street and Oxford Street.

The present Schuster Laboratory is on Brunswick Street, on the opposite side of Oxford Road. It has a historically oriented display in the entrance foyer, with a bust of Arthur Schuster. The most fascinating part for most visitors will be a display of photographs taken during Rutherford's time. There is Rutherford with Hans Geiger, for example, taken in the old laboratory — this is the Geiger who invented the famous radioactivity counter. A later group photo includes James Chadwick, who discovered the neutron, and Henry Moseley, the brilliant young chemical physicist who was killed in 1915 at the battle of Gallipoli (*see* OXFORD p.160). The building's lecture theatres are named after Rutherford, Moseley, William Bragg (x-ray crystallographer), and P.M.S. Blackett (famous for his work on the particulate nature of cosmic rays). All these people were here, students or associates of Rutherford — a roster without parallel in modern physics.

Chemistry and Engineering.   The foyer of the chemistry building has displays about current research in the department, without a historical component. When we visited in early 1994 there was a good display of historical posters and photographs, but it was a temporary exhibit, on loan from the Museum of Science and Industry. In a hallway off the foyer we can find a bust of Chaim Weizmann, who held appointments here from 1904 to 1917 and whose many accomplishments include novel procedures for the preparation of acetone, which made an invaluable contribution to the production of gunpowder in World War I. He is of course better known worldwide as an ardent Zionist who influenced the British government to promote the formation of a Jewish state — Weizmann became the country's first president when the state of Israel was ultimately proclaimed in 1948.

Another noteworthy historical figure is Osborne Reynolds (1842–1912). He spent his entire career here and represents a rarity in academia — an "academic engineer", exploring the physical foundations for processes that are used in industrial production. He is most famous for his work on the flow of fluids in pipes, where he identified the transition between laminar (smooth) and turbulent flows and defined the so-called Reynolds number, which relates this transition to the pipe diameter and the rate of flow — the subject is of great practical importance because onset of turbulence greatly increases resistance to flow. Reynolds is memorialised in the Engineering building, where the Thermodynamics laboratory is named after him. Within the laboratory there is an exhibit of pumps, turbines and other apparatus designed and used by Reynolds; in particular, a glass case displays the original apparatus used by him in 1883 to study the laminar/turbulent flow transition.

The Chemistry and Engineering buildings are on Brunswick Street, on the opposite side from the Schuster Laboratory. The Reynolds laboratory is a teaching laboratory and most of the space within it is taken up by large and intimidating prototypes of industrial devices. There was no class in progress when we visited and no difficulty in gaining access, but the situation could be different when there are classes going on. (It is worth noting that the Chemistry department has an individual in charge of displays, designed to present a more public image. More displays related to history may therefore appear soon.)

4. Salford and Barton

Joule House. James Joule's circumstances were quite different from those of John Dalton before him (who was independent, but poor) and Osborne Reynolds and others after him, who had University appointments. Joule's father was a prosperous brewer and set up laboratories for his son, at his own expense and (most of the time) in his own residential establishment. From 1843 to 1861 (which includes the period of the St. Ann's Church lecture and the work leading up to it) the principal working residence was at "Oak Field", on Upper Chorlton Road, about 2 miles south of Manchester city centre, but this house no longer exists. However, there was a hiatus (1849–1854, during which time Joule's first child was born) when James had his home in the formally distinct municipality of Salford (west side of Manchester) and set aside space for experiments there, although the laboratory at Oak Field was probably maintained as well. The Salford home is well preserved and bears an appropriate plaque; it is part of an attractive square of buildings (Acton Square) housing municipal offices. The neighbourhood was "posh" in Joule's time, the square gives one a good feel for the way it was.

Acton Square is off Windsor Crescent, close to Salford Crescent railway station, about 1.5 miles from Manchester city centre. (It is perhaps worth noting that Joule was born in Salford, on New Bailey Street, in an area in which everything historic has been obliterated by a bridge across the Irwell River.)

Barton Bridge.   The Bridgewater Canal (*see* p.236 above) was constructed at the behest of the the the Duke of Bridgewater, with engineering skills provided by James Brindley. One of the features that excited public admiration was the bridge that carried the canal over the river Irwell at Barton — one navigable stream crossing another was considered a sensational achievement at the time. The original aqueduct bridge had to be replaced around 1890, when the bed of the Irwell became part of the Manchester Ship Canal, but its successor is even more sensational, one of the marvels of Victorian engineering — a *swinging bridge*, which, filled with water, swings out of the way to allow tall ships to pass. The place is a tourist attraction, with benches and picnic areas from which to watch.

The aqueduct bridge is parallel to the Barton Road bridge across the Ship Canal, which also swings out to make room for ships. Barton Road itself turns off Liverpool Road (A57) just beyond Eccles, about 2 miles west of Salford.

Marston Moor (near York, North Yorkshire)

The battle of Marston Moor on 2 July 1644 was a disaster for the Royalist forces, who were crushingly defeated by Cromwell's army, with the loss of 4000 of their men. One who died there was the amateur astronomer William Gascoigne (1612–1644), who has earned his place in history not through specific telescopic observation, but by virtue of an inventive mind that hugely improved the precision with which telescopic measurements could be made. He noticed one day that a *single strand* of a spider's web in and around his telescope stood out in sharp focus. He realised that this strand happened to be at the focal plane of the telescope (i.e., where the image in an astronomical telescope becomes inverted) and that a measuring scale at the same position would also stand out sharply, permitting accurate estimate of relative dimensions or distances within the field of view. To supplement this, he went on to invent a micrometer for determining tiny displacements — the two inventions together converted astronomical measurement from an art into a science. Because of his early demise at the battle, Gascoigne's micrometer was all but forgotten, to be re-invented in France in the 1660s. Historians recognise his priority on the basis of letters preserved after his death; an official claim for priority on his behalf was made by the Royal Society in 1667.

Gascoigne had come from his home in the outskirts of Leeds to fight in the battle, the unachieved purpose of which was to keep York from falling into the hands of the enemy Roundheads. A tall roadside obelisk marks the place of battle and a tablet behind it describes the details. The dead, it tells us, were all buried in a mass grave immediately behind the monument. Commanders of units of the battling forces are named, but Gascoigne is not mentioned.

The battlefield and monument are on the north side of the minor road from Long Marston to Tockwith; Long Marston is on the B1224 between York and Wetherby. (See BURNLEY p.219 regarding preservation of Gascoigne's papers and micrometer.)

MARTON (near Middlesbrough, Cleveland)

Captain James Cook (_see_ WHITBY p.251) was born in 1728 in the village of Marton, now a suburb of the industrial city of Middlesbrough. The community has erected a stylish museum in Cook's memory at the site, in what is now called Stewart Park, a large green recreation area. The museum consists mainly of a succession of tableaux, illustrating Cook's early life, his voyages and his death. The fact that the first voyage, on _H.M.S. Endeavour_, had a scientific goal (observation of the transit of Venus across the sun), exploration being secondary, does not seem to be mentioned. However, there is a portrait of Joseph Banks, the naturalist who accompanied Cook on this voyage, above the reception desk. Also there is a cutaway model of the *Endeavour* showing how the internal space was arranged: Banks had what is called a "spacious cabin", 7 feet by 10 feet, without windows or ventilation. A far cry from Banks's huge homes on land (*see* REVESBY, Lincolnshire p.212) and a reminder of the risks and sacrifices men like him were willing to make for the sake of new knowledge and scientific understanding.

The actual site of Cook's birthplace cottage is marked by a granite urn, about 40 yards south of the museum entrance.

Visitors not coming directly from Middlesbrough will normally approach Marton via the A19 and A174, where signposts direct one to the museum. The museum is open all year, daily except Monday, 10–4.45. Winter hours are somewhat more limited. Phone (01642)-311211.

Great Ayton. In Great Ayton (4 miles south of Marton) the family cottage of Cook's boyhood is marked by an obelisk from Australia — the cottage itself was removed in 1934 and sent to Australia, where it now stands in Fitzroy Gardens, Melbourne. The old parish church in Great Ayton is worth a visit — Cook's mother and five of his brothers and sisters are buried in the churchyard.

MUCH HOOLE (near Preston, Lancashire)

Jeremiah Horrocks's observation of the transit of Venus across the sun in 1639 was a miracle of genius, coupled with some good luck (*see* ASTRONOMY p.25). How did the young curate of St.Michael's church manage to predict this rare event, when Kepler's tables failed to do so? What was the probability that the event would happen neatly in an interval between a string of Sunday services? His own account gives the spirit of it, telling how he looked at every opportunity all day, without at first seeing any unusual perturbation,

> ... in all these times I saw nothing on the Sun's face except one small and common spot, which I had seen on the preceding day and which also I after-

wards saw on some of the following days. But at 3h 15m in the afternoon, when I was again at liberty to continue my labours, ... I beheld a most agreeable sight, a *spot*, which had been the object of my most sanguine wishes, of an unusual size, and of a perfectly circular shape, just wholly entered upon the Sun's disk on the left side, so that the limbs of the *Sun* and *Venus* exactly coincided in the very point of contact. I was immediately sensible that this round spot was the planet *Venus*, and applied myself with the utmost care to prosecute my observations.

More than two hundred years later one Robert Brickell was appointed rector of St. Michael's and became fascinated with the Horrocks story. He wrote a small book entitled *A Chapter of Romance in Science*, and set about the task of providing a proper memorial for its hero. The results are there for us to see today; the church is full of references to the great event. The east window depicts the sun with the planetary circle upon it, and the artist's impression of Horrocks in the act of observation. A marble tablet tells us

St. Michael's Church has memorials for Horrocks inside and out. The porch dates from Horrocks's time, but all else (including the sundial) is relatively recent.

243

about Horrocks — "loving science much, he loved religion more". On the outside there is a large sun dial over the south door with the motto, *Sine sole sileo* ("Without the sun I am silent").

Horrocks lived at Carr House nearby and may have served as tutor for the owner's children — he had a south-facing first floor room directly over the porch, and that is where he actually made his observations. The house has been preserved almost unchanged in external appearance — still a grand sight. On the fateful Sunday we must imagine Horrocks shuttling back and forth between house and chapel, peering at the image in his camera obscura whenever the heavy schedule of his spiritual duties allowed it.

The church is at the south edge of Much Hoole, about half a mile from the southern turn-off to the village off the A59. It is kept locked when not in use. Phone (rectory): (01772)-612267. Carr House is at the road junction itself, a private residence not open to the public, but easily seen from the road.

NEWCASTLE (Tyne and Wear)

Newcastle was a coal mining centre from the middle ages onwards; for the last 200 years it has been a city of heavy industry — locomotives, steamships and the like.

Museum of Science and Engineering. The history of Newcastle's industrial growth is presented in an extensive and well-designed exhibit in this museum. Pride of place goes to George Stephenson (1781–1848), who was born just outside Newcastle (WYLAM, *see* p.254) and set up the factory in the city that built the "Rocket" and other famous early steam locomotives. His career is a prime example of the speedy application of scientific knowledge to transform the lives of people all over the world, though his own contribution to basic scientific knowledge was no longer comparable to what some of his predecessors had done — he was never a fellow of the Royal Society, for example. (The museum also has a gaudy "science" section, intended for children, with flashing lights and little intellectual content.)

Literary and Philosophical Society. Preeminence of engineering projects does not of course mean that interest in "pure" science was ever absent. The Newcastle Literary and Philosophical Society was founded in 1793, only 12 years after its more famous namesake in Manchester, and it still flourishes today. George Stephenson gave a demonstration here in 1815 of the miner's safety lamp he had invented, similar to and as good as the one invented independently by Humphry Davy — both men freely gave credit to each other when the coincidence became known.

Hancock Museum. This is another museum worth noting, with reasonably good exhibits on geology and natural history. It also has a small attractive room devoted to another local figure on the periphery of science, Thomas Bewick (1753–1828). He was an engraver of great note (his prints are still used on greeting cards today), with special interest in birds: he published one of the first popular guides to British birds, a rival in historical terms to James Audubon's later but better-known _Birds of America_.

The "Lit and Phil" is on Westgate Road, just east of Central Station. The Museum of Science and Industry (ominously in the process of being renamed "Newcastle Discovery", with the motto "It's all GO") is on Blandford Square, off Westmorland Road, west of the station. The Hancock Museum is adjacent to the University on Claremont Road. Both museums are open all year, Mon–Sat, 10–5. Hancock is also open Sun 2–5. Phone "Discovery" (0191)-232-6789; Hancock (0191)-222-7418.

Killingworth. From 1805 to 1823 Stephenson lived in Killingworth, 5 miles north of Newcastle, and all his early pioneering work was done there — the house is called "Dial Cottage" (it has a sundial over the door) and bears an appropriate inscription.

Dial Cottage is on Great Lime Road (B1505) west of Palmersville Metro station and the sports ground.

OUTHGILL (near Kirkby Stephen, Cumbria)

Come to Outhgill to learn about Michael Faraday's origins, just in case "born in London" should conjure up some mistaken notion of urban sophistication. Outhgill, in the rugged country of the northern Pennines, was a stopping place on what used to be a drovers' route for moving cattle or sheep to market. It had an inn and a smithy next door, where James Faraday was the blacksmith, earning his living off the passing trade. The Faradays left early in 1791 because hard times and a harsh winter had made it impossible for James to continue to sustain his family, and London presumably offered better security. We don't know the exact date of arrival in London, but we know the date James joined the Sandemanian meeting house there, which happened on 20 February. Michael Faraday was born on 22 September 1791, which raises an intriguing question. Was Mrs. Faraday pregnant when they started the journey? Was Michael conceived in Outhgill?

Outhgill seems to be on hard times again today. The inn is gone and most of the buildings stand empty, many of them are clearly deteriorating. There are no plaques or signs to tell the passer-by of the Faraday connection.

Outhgill is in the Eden Valley, on the B6259, about 5 miles south of Kirkby Stephen.

SCARBOROUGH (North Yorkshire)

The geologist William Smith (*see* HIGH LITTLETON, Avon p.126) came to this popular Yorkshire seaside resort for the last years of his life. He produced new local geological maps, he designed a museum — an architectural gem — for geological specimens and other objects of scientific interest, and he helped launch two young men onto distinguished scientific careers: William Crawford Williamson, the son of the museum's first keeper, who became noted for his work on fossil plants, and his own nephew John Phillips, who briefly turned the spotlight in the city of York onto science in 1831 (*see* YORK, p.254).

Rotunda Museum. The museum designed by Smith (subsequently expanded) retains only one item from its original geological collection, a frieze on the upper floor, completely surrounding the base of the Rotunda dome, depicting a geological section of the coast from the Humber to the Tyne. It was made by John Phillips and would have been particularly effective when the display cabinets below it were filled with geological finds and fossils, but these have since been moved to the Wood End museum; the cases now exhibit local crafts, such as embroidery work. The ground floor of the Rotunda has an archaeological exhibit, which includes a massive oak coffin found in a nearby tumulus in 1834, with a skeleton inside from around 1500 B.C. — its published description was written by the aforementioned W.C. Williamson, a precocious lad of 17 at the time.

Wood End, Natural History Museum.   This 19th century mansion used to belong to the Sitwells and one wing is devoted to the authoress, Edith Sitwell, who was born here. The remainder is a well-presented collection of indigenous birds, mammals, and fish, and a particularly good geological display. A distinctive feature is an emphasis on how local features relate to the geological history of the earth, with maps to direct the visitor to where the local features can be seen.

The Rotunda and Wood End museums are close to each other, one in the valley below St. Nicholas Cliff, the other on the Crescent above it. They are open Tue–Sat, 10–1 and 2–5; in summer also Sun, 2–5. Phone (Rotunda) (01723)-374839; Wood End (01723)-367326.

SELBY (North Yorkshire)

To have discovered and named two of the chemical elements is surely an extraordinary accomplishment and Selby native Smithson Tennant (1761–1815) did just that. The elements are iridum and osmium and were discovered by Tennant in his home laboratory in London. He had formed a business partnership with William Wollaston, to make and sell platimum boilers (for sulphuric acid production) and other platinum ware — the new elements are impurities in platinum ore that had to be removed to make malleable platinum metal. Wollaston, historically the better known of the partners, independently discovered two other contaminating noble metals, palladium and rhodium.

Tennant also discovered the chemical identity of diamond

with other forms of carbon: both yielded CO_2 on combustion, in exactly the same amount per gram. People didn't pay much attention — who wants to burn diamonds? Lavoisier made a similar experimental observation, but was unwilling to commit himself to a definite conclusion because diamonds are so unique in their crystal properties.

Tennant's father was nominally a clergyman in an Essex parish, but he was born and had his roots in Yorkshire; his wife came from an old Selby family (name of Daunt) and the habitual Tennant home appears to have been with them; Smithson himself lived here to the age of 20 and was educated in Yorkshire schools. In spite of this close association, Selby pays scant heed to Smithson's feats — civic pride rests on a much earlier era when the Benedictines held sway and built the superb Abbey Church in the town centre (where Smithson was baptised). A street named after Tennant, in a residential area on the north side of the town, is the only tangible reminder of his existence. A brief biography appears in the standard history of Selby, published in 1867.

Curiously, William Wollaston is also largely forgotten in his home town. (*See* EAST DEREHAM, Norfolk p.204).

THORNHILL (near Dewsbury, West Yorkshire)

John Michell (1724–1793) was another clergyman, this one with a modern flair for simple numerical calculations that brilliantly bring new ideas into a field of science — not necessarily *proving* the idea in question, but directing future experiments towards the goal of verification. Michell's calculations were in the field of astronomy. One calculation was statistical. He found that pairs of stars are seen apparently close to each other far more frequently than one would expect from a statistical calculation based on overall stellar distribution. This strongly suggested the existence of actual *double stars*, attracted to one another by physical forces, now known to be a common phenomenon. Another calculation was the first realistic estimate of the distance of a star from the earth. It was based on the apparent brightness of the star Vega relative to the planet Saturn (plus knowledge of the latter's relation to the sun). Let us assume for the calculation, Michell suggested, that the sun and Vega are equally bright at their surfaces — how much further away must Vega be to account for its relative dullness? The result was a phenomenal 460 000 times more than the distance of the earth from the sun! It actually turns out to be even more than that, because Vega is intrisically brighter than the sun, but that did not become known till 50 years later and Michell's estimate was in the interim a vital astronomical milestone.

Michell was rector of St. Michael's church in Thornhill from 1767 until his death. His astronomical work was published in the _Transactions_ of the Royal Society. Its relevance to the exercise of his duties (being, if you will, research into God's universe) would never have been questioned. He is buried at the church; there is a memorial tablet high on the north wall in the tower.

The church in Thornhill is on the direct road (B6117) from Dewsbury. It is kept locked between services. Phone (rectory) (01924)-465064.

WAKEFIELD (West Yorkshire)

This is the county town of West Yorkshire, long ago a centre of the cloth trade, but now eclipsed in size and industrial importance by the city of Leeds. A truly important figure in the history of British science, the clockmaker John Harrison (1693–1776), was born near here in the village of Foulby. What he did seems at first rather technical and lacking a purely scientific component — he devised a self-compensating clock pendulum and a mechanism whereby a clock can continue to run while being rewound — but in fact it led to the most accurate chronometer known up to his time, solving the "longitude problem" for which the Greenwich Observatory had been created nearly a century earlier (_see_ GREENWICH OBSERVATORY, London p.87) and earning Harrison a £20 000 prize, probably worth about a million pounds in today's currency.

The house in Foulby where Harrison was born is marked

by a plaque. Harrison's prize-winning chronometer, still in working order, is on display at the Greenwich Maritime Museum. The best we can do here is to see an early Harrison clock — with a mechanism made almost entirely of wood! — at nearby Nostell Priory, a stately home owned by the National Trust.

The village of Foulby and Nostell Priory (the latter well signposted) are about a mile apart on the A638, southeast from Wakefield. The Harrison house is on the highway, on the south side of the village. Nostell Priory is open from the end of March to end of October. From July to 8 September it is open daily except Fri, 12–5 (Sun 11–5), the rest of the year weekends only, Sat 12–5, Sun 11–5. Phone (01924)-863892.

Wakefield Museum. The city museum has a quaint exhibit devoted to Charles Waterton (1782–1865), who was born at Walton Hall, on the outskirts of Wakefield, and lived there, except when travelling, until his death. Waterton was an eccentric naturalist and explorer, whose book, *Wanderings in South America*, first published in 1825, was a national best-seller and one of the important catalysts for the period's burgeoning enthusiasm for amateur study of natural history. Waterton was a skilled taxidermist, who did not confine himself to preservation, but also created some purely artificial specimens from bits and pieces of real ones, sometimes just for fun and sometimes for political satire. The museum has a large collection, with both real and unreal specimens.

Wakefield Museum is open all year, Mon–Sat, 10.30–5, Sun 2.30–5. Phone (01924)-295351.

Walton Hall. At his Walton Hall home Waterton created one of the world's first dedicated nature preserves, where birds and mammals were protected from hunters, and trees and other plants were established to provide an attractive environment for them. The Georgian manor house (now a hotel) remains intact, exactly as pictured in sketches of Waterton's time, on an island in a small lake, with access via a neat iron footbridge. Much of the former estate has been encroached on by housing developments, but enough remains to retain the idyllic aura — wildfowl on the lake, herons roosting in the trees, etc. It is worth noting that Waterton's advocacy of the cause of preservation has received international recognition in Canada, where the Waterton Lakes National Park and the joint U.S.–Canadian Waterton–Glacier Peace Park are named after him.

The small town of Walton is on the B6378 between Crofton and Crigglestone. The turn-off for Walton Hall (a dead-end road) is signposted at the war memorial. The present name is Waterton Park Hotel, and the owners describe it as "a gem in a hidden valley". Phone (hotel) (01924)-257911.

WARRINGTON (Cheshire)

Warrington, midway between Liverpool and Manchester, is famous for its 18th century "dissenting academy", where Joseph Priestley taught and Robert Malthus was a student. Present county boundaries place it in Cheshire, and we have therefore described our visit to Warrington in the section on MIDLANDS (NORTH).

WHITBY (North Yorkshire)

Marton (near Middlesbrough), is where James Cook was born in 1728. Great Ayton, a few miles further south, is where he spent his boyhood and went to school. But the true heart of "Captain Cook Country" is Whitby, the port where Cook learned to love the sea and where the ships of his voyages were built — the celebrated *H.M.S. Endeavour* was originally a Whitby collier by the name of *Earl of Pembroke*.

It is not always appreciated that the primary mission of the *Endeavour*'s journey was scientific, namely to observe the transit of Venus across the sun, which was done in Tahiti. The better publicized mandate to settle the question of the existence of a southern continent was secondary and even that had its purely scientific aspect, a description of the fauna and flora of the putative continent and collection of specimens. A large party of scientists and assistants was on board, appointed by the Royal Society and headed by Joseph Banks, who later became notable on many counts, including the longest term ever as President of the Royal Society.

Cook himself made his own unique contribution to science by providing one of the first explicit testaments to the relation between diet and health, almost 20th-century-like in its emphasis on *vegetable* foods. On a trip to circumnavigate the world, the total travel time was three years and the remarkable and unprecedented medical fact about it was that not a single man was lost to scurvy! One died of a lingering illness, two were drowned and one was killed by a fall, but all the rest came home with health intact. Cook wrote a paper for the *Philosophical Transactions* to explain how he accomplished this by dint of careful attention to cleanliness and by his choice of what was fed to his crew. "Sour Krout" was a prominent component, "not only a wholesome vegetable food, but, in my judgment, highly antiscorbutic". Lemon and orange juice were mentioned as well, and "sweet-wort", a presumably alcoholic malt liquor. (Retrospective judgement would question the latter as a source of vitamin C.)

Cook learned the seafaring trade from John Walker, Whitby ship-owner, for whom he undertook a total of 16 voyages before volunteering for the Navy in 1755. When not at sea, he lived in an attic in Walker's house on Grape Lane, which is now the Captain Cook Memorial Museum. It exhibits authentic charts, manuscripts, furniture, etc., and it stands right at the waterside, where we can see across the river Esk to the shipyards where Cook's ships were built. One of the rooms in the museum is called the Scientist's Room, with an account of Joseph Banks's mission on Cook's first voyage and an admirable explanation of the transit of Venus across the face of the sun and what one learns from it. There are only a few specimens on display and nothing is said about scurvy.

Whitby also has an interesting general museum, situated in Pannett Park, with good exhibits of local discoveries — skeletons of marine lizards, erratic rocks from the ice ages, etc. There are displays devoted to Cook and to another local worthy, William Scoresby (1760–1829), an arctic explorer and versatile scientist who collaborated with James Joule on experiments into the magnetic side of the energy equation. This museum was founded in 1823 (at a different location) through the efforts of the Whitby Literary and Philosophical Society and the latter has a large library and meeting room at the museum, still in active use by its members. One other "must" for visitors is the bronze statue of Cook, on West Cliff, high above the town and harbour. There is a truly spectacular view from here, of the harbour and Whitby's famous Abbey and the open sea beyond.

The Captain Cook museum is open April to October, daily 9.45–5, also weekends in March and November, 11–3. Phone (tourist office) 01947-602674.   The Pannett Park museum is open all year, Mon and Tue, 10.30–1, Wed–Sat, 10.30–4, Sun 2–4. Weekdays in summer the hours are 9.30–5.30 daily. Phone (01947)-602908.

Staithes.   Whitby has fine walks and cliff scenery all around and we should not miss the opportunity to range up and down the coast a litle. The best place, perhaps, is Staithes, 10 miles north of Whitby, an ancient fishing village at the bottom of a deep ravine. It, too, is associated with Cook, for he had his first job here, working for a local general merchant before he moved to Whitby — the spot is marked by a plaque.

WIGTON (near Carlisle, Cumbria)

Here we have another instance of improbable origins. William Henry Bragg (1862–1942) was born at Stoneraise Place, a farm at Westward, near Wigton. The place seems rather remote, but apparently not enough so for his father, who sold the farm and all its furnishings in 1869 and moved himself and his sons to the Isle of Man. William entered the mainstream with a degree course at Cambridge, but then he too opted for far-away places, a faculty position at the University of Adelaide in Australia, which in 1886 must surely have seemed at least as remote as the Isle of Man. He did not return to England till 1908, by which time he was 46 years old. We shall refrain from detailing his movements after that — it suffices to say that he and his son (Australia-born William Lawrence Bragg) eventually entered the top ranks of British physics, the father as director of the Royal Institution in London and the son as Cavendish professor in

Cambridge. They shared the Nobel Prize for physics in 1915 for their development of the use of x-rays to determine the precise 3-dimensional structure of molecules in crystals, which (quite apart from its place in physics per se) is one of foundations of modern molecular biology (*see* CAMBRIDGE p.198).

Stoneraise Place remains as remote as ever and the present owners run it as a dairy farm. The Bragg family has installed a plaque above the front door of the farm house to record its distinguished place in history.

An unnumbered road to Westward is signposted on the A595 (just south of Wigton). This road ends at a T-junction; Stoneraise Place is reached by turning left and then left again. (Do not confuse with Stoneraise Farm, a different place in the same vicinity.)

WYLAM (Northumberland, near Newcastle)

George Stephenson (*see* NEWCASTLE p.244) was born in 1781 in Wylam, 8 miles west of Newcastle, in a small stone tenement built to accommodate four (clearly very poor) families. Stephenson had no schooling until he was 17, when he went to night school to learn to read and write. From then on, he was forever improving himself, taking steam engines apart, so he could learn how they worked and how to build better ones, etc. The Wylam house is a National Trust property and open to the public in summer.

Wylam and the Stephenson birthplace are 8 miles west of Newcastle, on the south side of the road to Carlisle (A69). The house is open from April to October, Thu, Sat, Sun and holidays, 1–5.30. Phone (01661)-85347.

YORK (North Yorkshire)

York is a city that everyone visits, to see the Minster, the city walls, the many archaeological remains, and the house where Guy Fawkes was born. The city was briefly in the spotlight scientifically in 1831, when the British Association for the Advancement of Science was created here and held its inaugural meeting at the splendid museum that had just been built by the members of the York Philosophical Society. The B.A. proved to be a viable and vital organisation during a period when the English scientific "establishment", then exclusively centred in London, Oxford and Cambridge, was smugly in the doldrums. (The Royal Society in 1830, faced with a choice for president of John Herschel or the Duke of Sussex, son of George III, elected the latter!) The B.A. held all its annual meetings in provincial towns and thereby enormously encouraged the latent research talents that existed outside the established centres. Why was York chosen for the inaugural meeting? Partly for the museum site already mentioned, and partly (very sensibly) because York is midway between London and Edinburgh — the latter city unaffected by the English blight.

John Phillips (1800–1874) was the principal organiser. Born in Wiltshire, he was orphaned early and had gone to live in Yorkshire with his uncle William Smith (_see_ SCARBOROUGH, p.246) and to assist him in his work. Phillips had become curator of the York museum by the time of the B.A. meeting. Later he moved to Oxford, where he was first keeper of the Ashmolean Museum and then founder of the University Museum. He was even professor of geology for a while. He died after a perhaps too hearty dinner at All Souls — he fell and slipped on the stone stairs as he was leaving.

The Yorkshire Museum today has an outstanding archaeological exhibit, but little else; it is open seven days a week all year, 10–4.30. A small observatory was built in the museum grounds following the 1831 B.A. meeting. It has a rotating roof designed by John Smeaton (_see_ LEEDS p.230); it is open on clear Thursday nights from November to February for public star-watching. Phone (museum) (01904)-629745. The Philosophical Society is still active, with an ambitious programme of public lectures; it is housed in the "Lodge" beside the main entrance to the Museum Gardens. Phone (01904)-656713.

8
Wales

BLAENAVON (Gwent)

Blaenavon, set in open countryside where the scars of former industrial activity have all but disappeared, was the scene (in 1876) of a truly significant research enterprise in metallurgical chemistry — and also the place where the results of that research were applied to profitable production of steel.

The problem is simple to state: making steel requires removal of impurities from iron ore. An old established method existed, but was prohibitively expensive. Invention of the Bessemer process in America reduced the cost, but it soon turned out that this process worked only for a minority of ores — it didn't work when phosphorus was a significant component. Many famous metallurgists had tried for several years to solve the problem (including the staff of the Siemens Company in Germany), but all failed until Sidney Gilchrist Thomas (1850–1885) tried his hand at it. Thomas was born in London and had been forced to abandon plans for a university education by his father's death. But he enrolled for evening lectures in chemistry while working at a menial job to earn his living, and his interest in steel-making is said to have been triggered by a remark of his teacher (named Challoner) that the man who could elminate phosphorus in the Bessemer steel-making process would one day make his fortune.

Thomas proposed on sound chemical grounds that silica

(SiO_2) should be able to displace phosphorus oxides from the molten iron. He was able to test this possibility with the help of his cousin, Percy Gilchrist, who happened to be analytical chemist at the Blaenavon works (which began operation in 1789), and who secured the cooperation of an unusually forward-looking works manager, E.P. Martin. The experiment succeeded and all those involved made their fortunes, as predicted, from patent royalties. (Thomas himself worked so hard at selling his process that his health failed and he died at the age of only 35 in Paris.) Today 90% of all steel is still made with the aid of the Thomas–Gilchrist method.

The iron works and ancillary buildings have been restored and are now a tourist attraction; an obelisk in memory of Thomas stands in the car park. Close by is a more popular tourist attraction, the Big Pit Mining Museum, where visitors have a chance to descend 300 feet in a miner's cage to a coal mine that ceased operating only in 1980. The scientific interest here is less, except perhaps in that it serves to remind us that iron ores are oxides that must be chemically reduced to produce the metal. Carbon is the usual reducing agent and coal mines and iron works are in close proximity all over the world for that reason.

Blaenavon is close to Abergavenny and 15 miles north of Newport. The Iron Works and Big Pit are clearly signposted and lie off the same service road. The Iron Works are open May to September, Mon–Sat 11–5, Sun 2–5, but guided tours can be arranged at other times. (One gets an excellent view of the buildings and some machinery from the outside even when the buildings are closed.) Phone (01495)-792615. Big Pit is open 9.30–5, seven days a week all year. Phone (01495)-790311.

CAMBRIAN ROCKS (Dolgellau, Gwynedd)

The textbook classification of rock strata from the Palaeozoic period (600–200 million years ago) is based on pioneering research in Wales and in the border areas of Shropshire; the names of the eras, *Cambrian, Ordovician and Silurian*, are based on the names of Wales itself and the ancient tribes who inhabited it (*see* GEOLOGY &

PALAEONTOLOGY p.33). We cannot quite travel in the footsteps of the pioneering geologists, see what they saw, and follow the exact clues that led them to their conclusions, because so much of the work depended on painstaking studies at home after rock samples were collected and fragmented. We can, however, identify areas which were their outdoor laboratories, little changed since the first geologists began mapping the area in the 1830s. Adam Sedgwick concentrated on the terrain here in North Wales and his colleague, Roderick Murchison, covered the border area, which is described in our section on MIDLANDS (NORTH).

Cader Idris. This mountain, at the southern end of Snowdonia National Park, was a focus for much of Sedgwick's research. The rock is of volcanic origin and was classified as Cambrian by Sedgwick, though it is now recognised to be Ordovician (500 million years old), a little younger than Cambrian (closer to 600 million years old). The mountain is (after Snowdon itself) the most popular venue for holidays and climbing in the National Park. A fine view is obtained from stopping places on the side of the A496, not far from the junction with the A470 running out of Dolgellau. In Barmouth, some 10 miles from Dolgellau, Panorama Walk above the Panorama Hotel affords another fine view of the mountain and the surrounding area. In both places the mountain's steep escarpment should be noted, ideal for sampling the rock over an enormous vertical distance and correspondingly large time frame of deposit.

Road to Trawsfynydd. The A470, running north from Dolgellau, passes through a forest area, popular for walks and other holiday activities. As the forest thins out, however, we begin to see low hills to the West, clearly very old, their tops flattened out. This is the core of the true Cambrian region, the oldest strata in which fossil remains of primitive living things can be found. Sketches of several of the peaks are found in Sedgwick's notes.

CARDIFF (South Glamorgan)

National Museum. The National Museum of Wales is a showpiece both for art and science. The outstanding exhibit in the context of science is called "Evolution of Wales" — it is of entirely new construction, filling a former courtyard, and was first opened to the public in October 1993. Its name is somewhat misleading because what it does is to teach the natural history and the geological history of the world and the evolution of the species that inhabit it, with only modest emphasis on Wales per se. The presentation is strictly sequential (it is difficult to "browse") and it is intended to teach the facts — this is the way it was (and is) — with little reference to how we know it all or to the scientists involved. It is also modern and high-tech: wide screen videos with loud running commentary appear at each successive stage of the tour, and there are some gimmicks, such as an animated mammoth. Nevertheless, within these limits of what some may see as undesirable features, the exhibit is accurate and complete. Concepts such as continental drift, ice ages, fossil types and their relation to evolution, are expertly explained; the museum is a good prelude for lay people to a trip through Wales, where so many of the places of scientific interest have to do with geology and palaeontology.

The National Museum has published a useful series of leaflets, "Geological Walks in Wales", giving precise directions to short walks and lucid explanations of what one can see there. The museum is in the heart of Cardiff and open Tue–Sat 10–5, Sun 2.30–5. Phone (01222)-397951 or 344111 (recorded information).

Llandaff Cathedral. This cathedral is the foremost in Wales. Its bishops (listed chronologically in the church) date back to the year 500. It was badly damaged by bombing in World War II. As part of its restoration, it was embellished by a modern Epstein sculpture — a giant arch and a figure of Christ — placed in a dominant position in the middle of the aisle.

The cathedral has a scientific connection through William Conybeare (1787–1857), who combined a career as a leading churchman with that of a respected geologist, one of the

260

early members of the Geological Society of London and a Fellow of the Royal Society. He was an able exponent of the middle ground between purely progressive theories of geological evolution and those that fell under the heading of "catastrophism" and thereby an influential voice in the days of Buckland and Lyell (*see* GEOLOGY & PALAEONTOLOGY p.36). His most important specific work was his detailed anatomical reconstruction of skeletons of two giant fossils, *Ichthyosaurus* and *Plesiosaurus*, from fragments discovered during this period, and their comparison with modern reptiles. Within the Church, he was Dean of Llandaff Cathedral from 1845 to 1857. His monument,in the graveyard on the south side of the cathedral, is a prominent obelisk: the inscription praises his many services to the church, but makes no mention of his scientific achievements.

Llandaff Cathedral is 2 miles northwest of the city centre, in the suburb of the same name.

CONWY (Gwynedd)

Conwy, with its castle and high town walls, is the most picturesque town on the north coast of Wales. It is also the site of one of the two difficult river crossings on the road to Holyhead, the traditional route to Ireland — the other such obstacle being the Menai Straits about 15 miles further west — and this makes for a splendid lesson in bridge construction, a monument *par excellence* to the best in Victorian engineering. (Another engineering monument of like quality is the dam at Lake Vyrnwy, *see* p.265 below.)

A design problem faced by the engineers was that bridges needed to be sufficiently high above the water to allow ships to pass below them. Also the populace had to be convinced that the structures would not be eyesores. The major scientific problem, however, was in the choice of materials — cast iron (high carbon content), as then customarily used for small arched bridges, has great resistance to compression, but is brittle and fatally weak in tension; on the other hand, wrought iron (low carbon content) could not bear the direct weights that cast iron can. A compromise was provided by suspension bridges, with beams made of cast iron and chains of wrought iron, but even these would not suffice to carry railway trains, vastly heavier than any horse-drawn vehicles. New techniques had to be devised and men like Robert Stephenson (the son of "steam engine" George Stephenson) had to muster all the genius of which they were capable. Empirical testing of the effects of shape and thickness on resistance to stresses had to be pursued with great intensity. One outcome of such work was the famous tubular bridge, made of hollow beams or girders — the research showing (perhaps surprisingly) that rectangular cross-sections gave the greatest strength, though the resulting box-like structures could not be called objects of beauty.

Here at Conwy we have three bridges side-by-side, almost touching each other, one end of each bridge anchored in the rock on which the castle stands. The first is the graceful suspension bridge, designed in 1824 by Thomas Telford, a man who was as much an architect as an engineer and did a brilliant job in matching the bridge

Stephenson's tubular bridge, completed in 1848. The trains ran through the "tube" and still do so today.

towers to the castle's architecture. This bridge was judged unsuitable for the railway and the adjacent tubular bridge was built by Stephenson and his associates to carry the trains — it was finished in 1848 and has carried trains uninterruptedly since then, and still does so today. The suspension bridge carried road traffic until 1958, when the last bridge in the trio (modern reinforced concrete) took its place, the suspension bridge remaining as a footbridge maintained by the National Trust. To make the history complete, we should add that even the road bridge has proved inadequate for 1990s through traffic to Holyhead and the ferries to Ireland: this traffic is now carried _under the river_ by a recently built tunnel, about a mile to the north. Fierce objections from conservationists and the local populace prevented it from being a fourth bridge, which would have dwarfed the others and the castle as well.

There were Telford and Stephenson bridges at Menai Straits as well, longer than those here and technically probably more difficult to design. However, they were not adjacent as here and are not preserved. An informative visitor centre adds to the attraction of the Conwy site.

CWM IDWAL (near Bangor, Gwynedd)

Twenty thousand years ago, almost yesterday in the history of the Earth and its inhabitants, much of the northern hemisphere was buried under huge sheets of ice, in places more than a mile thick. In Europe the ice extended from the North Pole to the shores of the Mediterranean. In America it covered the northern third of what is now the United States. In a few places (e.g. Switzerland) we can still see glaciers that are residues of this ice age, but in Britain we must be satisfied with evidence from rocks and earth to indicate their former presence. A nature trail at Cwm Idwal ("cwm" meaning valley or cirque), maintained by the Nature Conservancy Council, is an ideal spot to see this evidence, in a spectacular setting on the north side of Snowdonia. There is an historical association as well: Charles Darwin and Adam Sedgwick came here to see for themselves almost as soon as the Ice Age theory was first accepted by geologists around 1840. Darwin published an account of his visit in 1842.

The floor of the glacial cirque here is filled by a lake (Llyn Idwal); the nature trail, about 2 miles long over rough terrain, goes around the lake. Looking across the lake, we can easily imagine the whole cirque embedded in ice. From the top of the land (off the trail, just north of the starting point), we see the results of the erosive action of the ice, smooth and sloping upstream where rocks were torn away by the ice, steeper and scarred further downstream where moving rocks scraped at the valley walls. All around the lake and trail we see moraines, huge mounds of rocks left behind when the glacier finally receded and now, of course, covered with vegetation. Peat beds are seen about half way round the lake; we are told that pollen deposited within it provides a record of the gradual change in vegetation over the 10 000 years since the last permanent ice disappeared. Moraines and rock debris trail off towards the north along the A5, the direction of glacial flow. Here we know, of course, where they originated, but it must be appreciated that boulders could be transported for many miles, ending up in remote locations where they are geologically out of tune with their environment — there they become the so-called "erratic boulders", which constituted (worldwide) the first hard evidence for earlier ice ages. There are erratic boulders here, too, derived from unknown distant higher ground: one such, 6 to 10 feet high, stands close to the car park, where the footpath that leads up to the nature trail begins.

Cwm Idwal is reached by a half-mile footpath from the mountain school at Ogwen Cottage, where there are several buildings and a public car park — on the A5, 3.5 miles southeast from the village of Bethesda. A pamphlet by Ken Anderson, *The Ice Age in Cwm Idwal*, describes the area; it is likely to be available at the kiosk at Ogwen Cottage. The warden of the nature reserve can be contacted at Penrhos Road, Bangor, Gwynedd LL57 2LQ, phone (01248)-372333.

GLYN CEIRIOG (near Llangollen, Clwyd)

Wales has the world's best slate deposits and several slate mines or quarries are open to the public. The small Chwarel Wynne mine at Glyn Ceiriog is particularly good educational value. The mine is no longer in use, but there is a 30-minute guided tour of the underground caverns. A small museum contains hammers and chisels and other equipment used in the days of production. A video with commentary explains the principles and difficulties of slate mining and processing and also records the history of the bitter conflicts between mine owners and workers; it is a vivid record because much of the film footage comes from the 1920s. Guides and other staff are knowledgeable and explain how slate was formed around 500 million years ago. Huge underground pressures were the primary cause, but the mineral's unique cleavage planes, which give rise to the dense and waterproof sheets for which slate is famous, turn out to be *at right angles* to to the direction of the applied force. A good lesson in mineralogy!

Glyn Ceiriog is 3 miles south of Llangollen on an unmarked road, or it can be reached by taking the B4500 from Chirk on the A5. The mine is signposted in the village. It is open 10–5 daily, Easter to October. Phone (01691)-718343.

LAKE VYRNWY (Powys)

Lake Vyrnwy and the dam that creates it are another fascinating monument to Victorian engineering. This one dates from late in the Queen's reign: work was begun in 1881 and completed in 1892. The project arose to provide a reliable public water supply for the city of Liverpool, 50 miles away, which it did to the tune of about 50 million gallons a day, and it continues to do so now, more than 100 years later. The dam is 48 yards high and nearly 400 yards long; the road to the visitor centre and other buildings passes across it. Scenery is spectacular and bird life abundant — the Royal Society for the Protection of Birds is a co-sponsor of the visitor facilities.

Massive scientific and technical expertise went into the construction of the dam. The choice of the site is one example, based in part on the knowledge that this area has a huge annual rainfall: 1900 mm (75 inches) per year. Mostly, however, it was a confident understanding of soil mechanics, hydraulics, and of building materials and their response to erosion, pressure, etc. Twenty years earlier a dam constructed for the city of Sheffield had collapsed when the reservoir was first filled and 250 lives were lost! That had been an earth dam, as was normal then, but Vyrnwy was designed to be a masonry dam, the first of its kind in Britain. The Silurian rocks came from a quarry nearby and some of the individual stone blocks were enormous, weighing as much as 12 tons apiece. Hundreds of stone masons were employed to dress the stone and fit the pieces together. At the time it was built, the artficial lake it created was the largest in Europe.

Lake Vyrnwy is about 20 miles west of Welshpool and best approached from that direction, via the B4393, which goes on as a scenic drive around the entire lake. The visitor centre has a walk-through gallery with posters and recorded commentary. It is open daily, 10.30–5.30, April to December; weekends only at the same time from January to March. Phone (01691)-173278.

OSWESTRY (Shropshire, England)

Edward Lhuyd (1660–1709), a notable Welshman among Britain's "early lights", was born across the border in England, in Oswestry. However, he was illegitimate and was sent out to be nursed by one Catherine Bowen, in Crew Green (10 miles south of Oswestry), which is *within* Wales by present boundaries. (*See* Oswestry, Shropshire, for main entry for Lhuyd p.183.)

PRESELI HILLS (near Fishguard, Dyfed)

It is established beyond possible doubt that the Preseli Hills in Wales are the source of one of the stone types used to construct Stonehenge — the circle of smaller so-called bluestones, about 5 feet high. This means that these stones had to be transported over a distance of 200 miles, in contrast to the larger sarsen stones of the monument, which had a local origin. The question is, how were they transported? The dominant opinion is that they were carried by the builders themselves, first overland on sledges and rollers to Milford Haven, thence by boat to the English shore and up rivers to Salisbury Plain. One argument against that is a lack of convincing motivation — why go to all that trouble? — and another hypothesis is therefore currently gaining ground. This is that the stones were transported by glaciers, that they were common on Salisbury Plain as "erratic boulders" (_see_ CWM IDWAL p.263), and that to the builders of Stonehenge they were simply part of the readily available building materials. There are arguments against that, too: glaciation in the most recent ice age (20 000 years ago) didn't penetrate far enough south and a much older glacial period has to be invoked for the transport; there are no Bluestones dotting Salisbury Plain today.

Evidence from the Preseli Hills themselves tends to support glacial transport. No two rock outcroppings in the area have exactly the same chemical composition and careful anaylsis of Stonehenge samples can therefore pinpoint exactly where they came from. The results show that they are not from a single quarry, as they might be expected to be if purposely sought out by human agency; they seem instead to have a mixed origin, such as a glacial boulder flow might have.

One area of igneous rocks from which quite a number of Stonehenge bluestones are derived can be seen from an unnumbered road between Crymnich and Mynachlogddu. An ancient track running west from Llainbanal (1 mile from Crymnich) can take walkers closer to the actual rocks at an outcrop named Carnmenyn. This track, incidentally, is more than 3000 years old and is thought by archaeologists to have been part of a trade and communication route from western England to southern Ireland — i.e., movement of people over hundreds of miles took place, regardless of whether their baggage included rocks weighing many tons.

Another source of the bluestones (6 miles away as the crow flies) is the high-lying Carningli area, about a mile south of Newport.

Crymnich is on the A478, about 11 miles south of Cardigan. The turn-off to Mynachloggdu is at the south edge of the village. Newport (Dyfed), not to be confused with the town of the same name near Cardiff, is on the A487, 6.5 miles east of Fishguard.

SILURIAN ROCKS (Shropshire, England)

Although the Silures were a Welsh tribe, who lived in Wales at the time of the Romans, the prototypes of the rocks named after them are actually across the border in Shropshire, not far from Shrewsbury, which we have listed under Midlands (North).

ST. ASAPH (Clwyd)

Explorers of continents are driven by the same passion as strivers for scientific understanding. One whose name is known to everyone is H.M. Stanley (1841–1904), but it is not generally known that he was Welsh (born in Denbigh, 5 miles south of St. Asaph) and that his name at birth wasn't Stanley at all, but John Rowlands. It's a most unlikely success story, for he was placed in a workhouse in St. Asaph at the age of six. He ran away from there at age 15, sailed as a cabin boy to America, was adopted by one Henry Morton Stanley (whose name he took), and became a well-known descriptive travel writer, whose destinations included esoteric places like Tibet and Ethiopia. An American newspaper commissioned him to go to Africa on the legendary search for the long-lost David Livingstone, and it was after that that he began his serious explorations, tracing the course of the Congo River, for example. He eventually returned to Britain to live and received a knighthood in 1899.

St. Asaph has a population of only a little over 3000, but is formally a city because it has a charming cathedral (fittingly also small), whose history goes back to the 6th century. The cathedral has a fine memorial to Stanley on the south side, made of the best Welsh slate. The old workhouse survives, now part of the local hospital (called the H.M.

Stanley Hospital), and there is a bas-relief of Stanley himself on the outside wall.

St. Asaph is on the main Chester to Bangor road (A55), about 30 miles from Chester.

USK (Gwent)

Alfred Russel Wallace (1823–1913), the co-proponent with Charles Darwin of the theory of the origin of species by natural selection (*see* EVOLUTION & ANTHROPOLOGY p.37) was born in Usk. He was the eighth of nine children in a family beset by economic problems; they had moved across the Severn from England to find a more inexpensive place to live — a place called Kensington Cottage, on the road to Llanbadock and Llangybi, facing the river Usk, with the high banks of the valley behind it. Wallace moved away to Hertford at age 5 to go to school, but he always remembered the rural surroundings here and the carefree days of his early childhood. He came back for a nostalgic visit at the age of 60, which he describes in his autobiography: he found "nothing changed between the bridge and Llanbadock; not a

Kensington House in Usk, where Alfred Wallace was born in 1823.

new house had been built and our cottage and garden remained as I had in memory." A photograph accompanies the text.

Surprisingly, not much has changed even now. Kensington Cottage has become Kensington House and is quite a bit grander than it used to be, but it still has an attractive garden, and the river in front and the wooded banks behind. A service station has been put up at the bridge (at the edge of the town), where the Llanbadock road begins, but the land in between is occupied by the playing fields of the Usk rugby football club, still green and uncluttered by houses. There is no plaque or other memorial at Kensington House, but there is a wooden bench in Wallace's memory in front of the Gwent Rural Life Museum, on New Market Street, close to the bridge, but on the town side of the river.

Usk is about 10 miles north of Newport, near the junction of the A449 and A472. The river runs south from the town; Kensington House is about 400 yards from the bridge; the road is signposted for Llangybi.

WREXHAM (Clwyd)

Institutions can have curious beginnings. Yale University (New Haven, Connecticut), one of the premier centres for science in the United States, is a good example: its origins are in northeast Wales; it is named for Elihu Yale, member of a family with much property hereabouts — there is a place still called Plas-yn-Yale about 12 miles to the west of Wrexham. It was actually David Yale, Elihu's father, who once lived in New Haven, but he was long dead when (in 1718) a small struggling school there desperately sought new benefactors. Elihu Yale, grown rich in the East India trade, responded generously and the college adopted his name in gratitude. Whether it mattered much to Elihu is not known; he had been born in Boston, but his family had returned to Britain shortly thereafter, and he had probably never even visited New Haven. He died in London in 1721 and was buried at St. Giles Church in Wrexham. His tomb is in the churchyard on the west side of the church and bears a curious (and long) inscription — "Much good, some ill he did," it states in part, "so hope all's even and that his soul thro' mercy's gone to Heaven."

St.Giles Church (signposted) is in the town centre, just off the High Street. Telephone (rectory) (01978)-263905 (or 355808).

9
Scotland (Mainland)

ABERDEEN (Grampian)

Aberdeen has two venerable institutions of higher learning, King's College (founded 1495) and Marischal College (founded 1593), which merged in 1860 to become the University of Aberdeen. But the city is famous today as the nucleus of Britain's oil industry, and seems to have little concern for its past association with science and technology. One such association is with James Clerk Maxwell, who taught at Marischal from 1856 to 1860 and found his wife here — Katherine Dewar, daughter of the principal. Another is with Arthur Keith (1866–1955), anatomist and anthropologist, who was born in Aberdeen and educated at the university. He then spent most of his career as conservator of the museum at the Royal College of Surgeons in London, where he established himself as a leading authority on human fossils. One of Keith's claims to fame (or notoriety) is his involvement in the Piltdown fraud — he was one of the expert judges of the unearthed skeleton, but the most recent book on the subject (F. Spencer, *Piltdown. A Scientific Forgery*, Oxford University Press) accuses him of having buried the skeleton in the first place!

Marischal College has a museum, half of which is devoted to anthropology, but it deals with the subject on a different plane — "the beauty and pain of the human condition" — with scant attention to scientific or historic matters and no mention of Keith at all. Another room in the museum, an A to Z listing of famous Scots, has a small display about Maxwell.

Marischal College today is a grandiose bulding (built 1905) on Broad Street; the museum entrance is within its central quadrangle. It is open Mon–Fri 10–5, Sun 2–5, closed Sat. Phone (01224)-273131.

BO'NESS (Central)

The Kinneil estate, now a fine public park, is the place where James Watt first tried (unsuccessfully) to produce a working steam engine to his new design — the one that still looked and worked like a Newcomen engine, but was much more efficient because it had a separate condenser to cool down the steam. The historic mansion was leased at the time to industrialist John Roebuck, who built a workshop for Watt next to a burn providing plentiful water. Unfortunately, the machinists at Roebuck's existing plant lacked the skill to construct the engine to the precise tolerances that were needed, Roebuck became bankrupt and had to sell his investment in Watt's patent to Matthew Boulton in Birmingham. The rest is history — Boulton and Watt transformed the world.

Watt's workshop cottage has been preserved and a "typical" steam-generating cylinder has been set up alongside. There is a fine museum on the grounds, with displays related to Watt and to the Roman relics that have been excavated in the area. Kinneil House itself is not in good repair and is closed to the public.

James Watt's workshop cottage on the Kinneil estate.

Bo'ness is about 10 miles east of Falkirk. Kinneil House and Museum are on the west side of the town, clearly signposted on the Edinburgh–Falkirk road (A904). The cottage and steam cylinder are in the grounds, which are always open. The museum is open April to September, Mon–Fri 10–12.30 and 1.30–5 and Sat 10–5 (May to August also Suń 10–5); October to March, Sat only, 10–5. Phone (01506)-824318.

CORSOCK and PARTON (near Dumfries, Dumfries and Galloway)

James Clerk Maxwell (1831–1879), though lacking a title, was born into the landed gentry and never lacked the means for a comfortable life. He was actually not a Maxwell at all, but belonged to the Clerk family, from Penicuik near Edinburgh — his father, John Clerk, had to adopt the surname of Maxwell (for complex legal reasons) when he inherited an estate in "Maxwell territory", near Dumfries, and took up his residence there.   Glenlair is where Maxwell grew up and it remained his home for most of his life. Even when he was on the faculty of King's College in London (where he developed his initial ideas on electromagnetism), he invariably spent at least four months of each year at Glenlair. When Maxwell died of cancer in 1879, at the early age of 48, he was in Cambridge, but his body was returned for burial to the Scotland he loved.

Glenlair.   This is one of the most scenic places in this book — green pasture land all around, patches of trees, scattered farms, all on rolling hilly terrain, with many picturesque views of the Urr River or Loch Ken below us. The Maxwell estate covered about 6000 acres, including tenant farms. Glenlair itself was built according to the design of Maxwell's father, a grand stone mansion by modern standards though relatively modest for the times. The house sadly was gutted by fire in 1929 and has never been repaired. The walls are still there, though, almost to the roof top, enough to recreate for today's visitor the bygone manorial atmosphere. The family that now owns the home farm of the original estate lives in the former servant quarters just a few steps from the ruins. They maintain the access lane, but do nothing about the main house itself and there appear to be no plans to preserve it nor to convert it to some modern habitation.

Parton. Glenlair is about five miles from the village of Parton, where the Maxwells regularly attended church. Maxwell's remains were moved to Parton after he died in Cambridge, and he is buried in the churchyard of Parton Kirk, in the ruins of an old chapel (the "Old Kirk"). The polished granite tombstone lists Maxwell's father and mother and his widow as buried in the same place. A brass plate was installed at the entrance to the churchyard in June 1989 — it calls Maxwell "a good man, full of humour and wisdom". A booklet published in 1979 by the community, as a tribute to Maxwell on the centenary of his death, is for sale at the Parton post office.

Corsock. The village of Corsock is even closer to Glenlair than Parton, and the church here, too, can claim close connections with the Maxwells. It was built in 1839 through the influence and exertion of Maxwell's father and both father and son were among the church elders. The church contains an impressive stained glass window, installed in Maxwell's honour. The motif of the window is the Magi following a brilliant Star of Bethlehem and a Greek inscription is embedded in the glass:

ΠΑΣΑ.ΔΟΣΙΣ.ΑΓΑΘΗ.ΚΑΙ.ΠΑΝ.ΔΩΠΗΜΑ.ΤΕΛΕΙΟΝ.

Loosely translated, it means "All good giving and every perfect gift" — undoubtedly a reference to the Epistle of James from the New Testament "All good giving and every perfect gift comes from above, *from the Father of the lights of heaven.*" A plaque placed next to the window describes Maxwell as "a genius that discovered the kinship between electricity and light and was led through the mystery of Nature to the fuller knowledge of God."

The above places are about 10 miles west of Dumfries, on roads that turn off the A75: Corsock on the A712, Parton on the A713;

Glenlair on a lane off the B794, signposted to Nether Corsock. The church at Corsock is normally kept locked, but keys can be obtained from an elder; ask at the village shop. (A narrow single track road can be used to go directly from Parton to Glenlair, considerably shortening the total distance for the Parton–Glenlair–Corsock circuit.)

CROMARTY (Highland)

Cromarty is an ancient fishing town, off the beaten path, seemingly forgotten by progress. The one anomaly is the sight of oil-drilling platforms in Cromarty Firth — their presence is due to a refitting and reconstruction facility at Nigg, on the opposite shore of the Firth.

The town's most prominent native is Hugh Miller (1802–1856), amateur geologist and fervent churchman. He began life as a stonemason, but also wrote poetry and essays. One of his hobbies was collecting fossils (especially fossil fishes) from local rocks, notably from what has become known as "Old Red Sandstone", a deposit that corresponds in age to the Devonian stone of Southern England, but is richer in fossils. Miller's fame as an amateur geologist spread rapidly, leading to associations with Roderick Murchison, Adam Sedgwick, and even the author, Thomas Carlyle.

Miller eventually moved to Edinburgh to edit *The Witness*, a church newspaper, and soon expanded its contents to include current affairs and to make it the second largest paper in Scotland. He combined his interest in geology with his strong religious feelings in a book *Footprints of the Creator*, which was in part polemic, a response to evolutionary ideas proposed by another Scotsman (and publisher), Robert Chambers (*see* PEEBLES, p.297). Though Miller never moved to London or became a full time geologist, his publications about red sandstone were highly influential in properly ordering the geological periods of the palaeozoic era.

Cromarty celebrates Hugh Miller with enthusiasm. A statue mounted on an enormous column, on a hill behind Miller's birthplace, dominates the town. The birthplace itself is Scottish National Trust property and has been converted to a museum. It retains its original thatched roof and

Welcome to Cromarty and the house where Hugh Miller was born.

has preseved the living quarters more or less the way they were, furnished with authentic period items. Upstairs are showcases with appropriate fossils and accurate explanations of what they tell us. Particularly fascinating are copies of correspondence with all the great and the good of the day. Well worth worth a visit!

Cromarty is 26 miles northeast of Inverness, near the tip of the Black Isle. Hugh Miller's Cottage is a National Trust property, open Easter to September, Mon–Sat 10–1 and 2–5, Sun 2–5. Phone: (01381)-600245.

DARVEL (near Kilmarnock, Strathclyde)

Alexander Fleming, one of the great modern benefactors of humankind through his discovery of penicillin, was born on a moorland farm near here. Every day he trudged 4 miles to go to school in Darvel; later he attended Kilmarnock Academy; then he was off to London, for a lifetime association with St. Mary's Hospital Medical School, which now has a museum that is dedicated entirely to Fleming and penicillin (*see* LONDON p.85). One never ceases to marvel at the inborn drive for knowledge and understanding that can produce a man of Fleming's stature in such isolated surroundings — even Darvel itself (at least as seen today) is a plain and uninspiring little town. It has a bust of Fleming at one end of a small garden built around the war memorial, but the inscription is badly worn. The Carnegie Library across the street has a framed picture.

Irvine.   Irvine, the modern industrial city on the opposite side of Kilmarnock, is the antithesis of Darvel and perhaps even of greater interest in relation to Fleming because it contains a huge factory of the Smith Kline Beecham pharmaceutical company, housing one of the largest bulk penicillin manufacturing operations in the world. Was it a deliberate tribute to Fleming, to place the plant so close to his place of birth? Apparently not — the glossy brochure put out by the company for public relations has a photograph of Fleming on page 1, but calls the factory location a coincidence.

Darvel is east of Kilmarnock on the A71. The Irvine factory is on Shewalton Road, close to the junction of the A78 and A71. Penicillin and related products are produced only in bulk and shipped from here to other plants all over the world for packaging and sale.

EAST KILBRIDE (near Glasgow, Strathclyde)

Two famous anatomists, the brothers William Hunter

(1718–1783) and John Hunter (1728–1793), were born on a farm, Long Calderwood, near East Kilbride, on the south-eastern outskirts of present-day Glasgow. The brothers advanced the course of anatomy by a combination of research, private teaching establishments, and medical prac-tice and grew wealthy in the process. John, the younger brother, leaned towards anatomical teaching and specimen collection. William Jenner, the "vaccinator" (*see* HUMAN BIOLOGY & MEDICINE p.8) was one of his pupils and the Hunterian Museum at the The Royal College of Surgeons in London (*see* LONDON p.62) is his great tanglible legacy to posterity. The older brother, William, had a rival anatomy school in London, but his greatest fame came from his own practical skills; he was the attending physician at Queen Charlotte's first confinement. William, too, founded a mu-seum in London, but in his will he left its contents and an additional sum of money to the University of Glasgow — that is the foundation of the university's Hunterian museum (*see* p.289). Microscopes and other souvenirs of the benefactor are on display in the museum.

After World War II East Kilbride became one of the deliberately planned New Towns, but the Hunter birthplace and an adjacent barn are preserved within the urban devel-opment and the house bears an appropriate plaque. The site has recently been acquired by the District Council and money has been appropriated to create a museum there, to be opened to visitors within the next 2 or 3 years.

The site is on Maxwellton Road in the district of Calderwood, directly opposite what are known as the "Calderwood Shops". Phone (01355)-220046 (local history department at the public library).

EDINBURGH (Lothian)

Edinburgh was at the the centre of the Scottish En-lightenment, a remarkable period in the 18th century, glori-fied by the presence of David Hume, the philosopher; Adam Smith, the founder of scientific economics; the chemist Joseph Black; the geologist James Hutton and his literate spokesman, John Playfair. "A hotbed of genius", it has been called. But this period, though exceptional, was by no means unique. Edinburgh can claim a distinguished scientific tradition long before the Enlightenment and afterwards as well: John Napier, inventor of logarithms, as early as 1600; the physiologist Robert Whytt around 1750; physician Joseph Lister in the 19th century. And James Clerk Maxwell, one of the stars of British physics, was born here in 1831 and went to school in the city. Reminders of all of these persons and their work can be seen throughout the city — houses where they lived, gravestones, museum displays.

Charlotte Square.   Joseph Lister (*see* p.284) lived after 1870 in a town house at 9 Charlotte Square, at the west end of George Street, in the heart of Edinburgh's "New Town". The house is intact and Lister's residence is commemorated by a plaque. Alexander Graham Bell (inventor of the telephone) lived here too, an Edinburgh product, though he moved to America in middle age. He was born at 16 South Charlotte Street, just off the square, where there is a plaque. New Town in general and Charlotte Square in particular are areas of beautiful and dignified urban design — fine residences for the affluent in the Georgian and Victorian years, worth a stroll today at the end of a busy day.

Greyfriars Church.   At the south end of George IV Bridge, close to the Royal Scottish Museum, is a sacred spot in Scottish religious history, Greyfriars Church. Here in 1638 the National Covenant was signed to uphold the Presbyterian form of worship. The Covenant was later renounced by Charles II and many bloody rebellions followed. Twelve hundred Covenanters were taken prisoner in a famous battle and then were kept penned up for five months in Greyfriars Churchyard. James Hutton is buried here within the area of the former covenanters' prison, at the south corner of the cemetery. The grave was unmarked when located in recent times, but a stone was placed in the wall on the occasion of the 150th anniversary of Hutton's death (1947), with the inscription, "Father of Modern Geology".

Physiologist Robert Whytt (1714-1766) is buried in the church itself, his grave marked by a handsome monument.

As we have not mentioned him elsewhere in this book, we note here that he was a personage of the first rank, whose work heavily influenced the heated debates that went on throughout Europe on the subject of "animism": is there a "soul" in the physical sense, a central basis for all the activities of life? Whytt contributed to the subject by demonstrating that muscle activity in animals can continue long after death, and by studies of reflex action, the origin of which he was able to confine to just a portion of the spinal cord. His book, *The Vital and Other Involuntary Motions of Animals*, is considered a classic in neurophysiology. On another tack, early in his career, Whytt studied the efficacy of lime water in breaking up bladder stones. This led the chemist Joseph Black (then in Glasgow) to investigate a whole series of alkaline substances as possible agents for curing "the stone" — it was in the course of this work that Black discovered the first known gas that is different from ordinary air (CO_2), which in turn led Joseph Priestley to his broad researches on all kinds of "airs" and the discovery of oxygen.

Greyfriars church and churchyard are closed on Saturday afternoons.

Maxwell in Edinburgh. Maxwell attended Edinburgh Academy (a school still active today). He was even born in Edinburgh because his parents had come here to assure the best possible medical care during his mother's confinement — she was 40 years old and an earlier child had survived only a few months. Those who want to follow in Maxwell's Edinburgh footsteps can see the house where he was born on India Street (marked by a plaque). Maxwell lodged at 31 Heriot Row, just around the corner, while he was attending school. Edinburgh Academy itself is about half a mile further north, on Henderson Row, an unmistakable aura of stern tradition emanating from its ancient buildings.

Merchiston Castle. The official residence of the Lairds was Merchiston Castle, which is now part of Napier University on Colinton Road, a little over a mile south of the city centre. The college is a large modern concrete structure built in 1964, but the castle was left intact and has been most tastefully incorporated into the modern buildings around it.

Merchiston Castle at Napier University.

The old gateway to the grounds of the castle has also been left intact. It will be noted that the "castle" is quite small and John Napier didn't actually live here until after his father died and he himself succeeded to the title in 1608 — two families in the building would have been a crowd. The invention of logarithms thus began at an earlier Napier home, but the first full publication came long after he moved to the castle, and it was to this castle that Henry Briggs came from London in 1615 and 1616 to discuss his decimal logarithms with Napier (*see* WORKING WITH NUMBERS p.14). (We visited the site on a Sunday, and the combination of old and new was perhaps particularly attractive because it was also tranquil — there were no students in evidence.)

Royal Infirmary. The former surgical hospital, where Joseph Lister worked and introduced antiseptic practices into surgery, used to be part of the Royal Infirmary, but is now a classroom and office building of Edinburgh University. It is a dour grey stone building, typical of much of Edinburgh, at the foot of Infirmary Street, right across South Bridge Street from the Royal Museum and the "Old

Plaque in honour of Joseph Lister and his mentor and father-in-law, James Syme, outside the Old Infirmary.

College" of the university. There is a plaque to indicate that James Syme (Lister's father-in-law) and then Lister "had charge of this building" between 1833 and 1877.

The present Royal Infirmary surgical wards are in a fascinating building of Scottish Baronial style on Lauriston Place, a few hundred yards to the west. The vestibule walls are covered by tablets recording royal visits and the names of donors who gave substantial gifts. Some of the donors go back to the 1700s, their names having been transferred from the Old Infirmary in 1881.

Royal Museum of Scotland. The Royal Museum of Scotland on Chambers Street is a national institution, housed in one of the finest Victorian buildings in the country. It is a comprehensive museum, serving both the arts and the sciences — the sculptured heads above the entrance (Queen Victoria, Prince Albert, James Watt, Charles Darwin, Michelangelo and Isaac Netwon) symbolise its origin and its scope. The spacious main hall is a splendid example of Victorian design, with fountains, ponds, and comfortable seating, which sets a note of cheerful enjoyment for the museum as a whole. The museum was originally opened as "The Industrial Museum of Scotland", and the scientific exhibits still tend to emphasise technology. But there is a good exhibit on evolution and also a fine "zoo" of large stuffed mammals.

From the Enlightenment period the museum has a collection of chemical glassware used by Joseph Black (1728–1799), one of the founders of modern chemistry, but not as well known as some of his contemporaries — Priestley and Lavoisier, for example — because he published very little. (But he was no recluse, so that the results of his work were widely disseminated by word of mouth and in letters.) Black was a brilliant and influential lecturer (10 years in Glasgow, 30 in Edinburgh) and most of what we know about his work today is derived from student lecture notes that have survived. Outstanding contributions were: (1) The discovery and study of "fixed air", obtained by heating or acidifying certain salts and found to be unable to support combustion — the salts were carbonates and we now know the gas as carbon dioxide, CO_2. (2) The discovery and the actual naming of "latent heat", i.e., heat that is not used to raise the temperature but is instead taken up or given off in phase transitions, which occur (as we all now know) at fixed temperature. The conceptual novelty of both of these items was astonishing; Black's influence on the chemical community was immense. James Watt is said to have been one of those who were influenced — latent heat is central to understanding how a steam engine works.

The museum is open Mon–Sat 10–5, Sun 2–5. Phone (0131)-225-7534.

St. Cuthbert Church. There is a formal memorial to John Napier in the centre of Edinburgh, in the vestibule of the former West Kirk (now Church of St. Cuthbert) at the west end of Princes Street Gardens. It is quite modest, but in a way a singular tribute because the church is small and the congregation must walk right by the memorial at every service.

Salisbury Crags and Arthur's Seat. James Hutton lived on St. John's Hill, on the eastern side of the present Edinburgh University campus. It is instructive to walk around here for a few minutes, and to see the good view one gets of Salisbury Crags, a spectacular line of cliffs at the edge of Arthur's Seat, an 800 foot ancient volcano (erupted 325 million years ago) lying within Holyrood Park. "Geology" was always there for Hutton to look upon, practically in his own backyard.

Hutton and his friends unambiguously identified intrusions of molten rock at Salisbury Crags, pushed into the predominant (and softer) sedimentary rocks, not lying passively at the base, where the primeval hard rock was supposed to be — he called such intrusions "uncomformities" and they became the basis for his cyclic theory of earth history (*see* GEOLOGY & PALAEONTOLOGY p.31). It is possible to drive along the perimeter of the crags (Queens Drive) and get a general view, but it is better to climb on foot to Radical Road, a path at the base of the upper cliffs. The best part of the cliff for observation is directly across from the Park Road gate to Holyrood Park and it is known as "Hutton's Section".

(A little to the north of Salisbury Crags is Carlton Hill, climbed by steps from Waterloo Place. There is a monument on top to John Playfair, whose concise account of Hutton's _Theory of the Earth_, published after Hutton's death, was the vehicle by which Hutton's ideas were widely disseminated.)

A geological "excursion guide" to the Lothian area, which includes Arthur's Seat, can be purchased from the Edinburgh Geological Society, c/o Grant Institute of Geology, West Mains Road, Edinburgh EH9 3JW.

GLASGOW (Strathclyde)

Glasgow abounds with "Kelvin this" and "Kelvin that", but they are generally named for the Kelvin River, not for Glasgow's premier scientific citizen, the thermodynamicist Lord Kelvin, whose remarkable career was described in our introductory text (_see_ HEAT & THERMODYNAMICS p.22). On the contrary, His Lordship, too, took his name from the river. He was born William Thomson, but when Queen Victoria raised him to the peerage in 1892, he proudly chose the name "Kelvin" for his title, after the stream that runs alongside the university campus.

Lord Kelvin is unique among the well-known scientists who have been historically asociated with Glasgow or its environs, in that he belongs wholly to Glasgow. For him it is not a case of being a Glasgow boy who went out into the world and made good and is therefore remembered with pride — he made good in Glasgow itself and stayed there. He went to Cambridge for his education, but returned home after his degree, became professor of natural philosophy at the University of Glasgow at the early age of 22, and held that position for an incredible term of 54 years, despite numerous offers from more prestigious institutions. He built himself a fancy home on the coast in Largs (mostly for summer use), which is where he died in 1907. Only after death was Glasgow spurned — Lord Kelvin is buried among the elite in Westminster Abbey.

Glasgow University. The university (Britain's fourth oldest) was founded in 1451, but moved to its present site only in 1870, during Kelvin's lifetime. The principal building is a grandiose Victorian Gothic edifice on the south side of

University Avenue, and includes two quadrangles, lecture rooms, university offices, etc., as well as the Hunterian Museum.

The old Natural Philosophy building nearby (now prosaically renamed "Physics and Astronomy") is where Kelvin taught and had his laboratory. It includes the Kelvin Lecture Theatre, a grand old hall which was indeed named for the man and not the river and contains a fine photograph of him in old age. A modern addition to the building has some Kelvin apparatus on exhibit and there is more in the "Kelvin Museum", which is used for conferences and not normally open to the public. Between the physics building and the Victorian edifice is Professor's Square, former housing for distinguished faculty. Kelvin lived in number 11, an enormous residence, which he continued to occupy, when in Glasgow, even after he built his own palatial home in Largs. It must have been expensive to heat — professors were expected to rent out rooms to students to help pay for this and other maintenance costs.

A Glasgow University student who subsequently did well (and who had attended Kelvin's lectures at the university) was the chemist William Ramsay, celebrated professor at University College (London), discoverer of most of the inert gas elements, winner of the Nobel Prize for chemistry in 1904. It is said that his interest in inert gases was aroused by a lecture given in Glasgow in 1875 by Norman Lockyer on the solar spectrum — Lockyer would of course have referred to helium, which he had identified as a solar element (not thought to be present on earth) on the basis of spectral lines.

Another celebrated chemist was Frederick Soddy (1877–1956), who was lecturer in physical chemistry at the university from 1904–1914. He invented the word "isotope" during his Glasgow residence — more than that, he had a clearer idea than anyone else of the *concept*, i.e. of the underlying reason for the existence of isotopes in terms of the composition of atomic nuclei. Soddy had worked with Rutherford and Ramsay before he came to Glasgow — he chemically confirmed helium as a product of radioactive decay in the course of this work — and thus was well prepared for providing the bridge he built between physics and chemistry in the subatomic field. He won a Nobel Prize for his work in 1921. A bronze plaque for Soddy was unveiled in 1958 at the Glagow meeting of the British Association.

Hunterian Museum.   On the second floor of the main university building is the Hunterian Museum, named after William Hunter, the famous anatomist and physician born in East Kilbride nearby (*see* p.279), who was a Glasgow student in the 1730s and in effect created the museum by bequests he made in his will. The museum today includes a fine art gallery in a separate building and in its main galleries emphasizes geology, archaeology and coins. However, it also has small collections of objects and pictures related to several Scottish scientists, including Lord Kelvin. A showcase in the general exhibit hall is devoted to Kelvin and there are two smaller cases on the stairs leading to the museum.   They show photographs of Kelvin's laboratories and some of the instruments he designed or invented, such as the marine compass (which contributed greatly to Kelvin's wealth), sensitive electrometers, etc.

The museum is open all day Mon–Fri, mornings only on Sat. Phone (0141)-330-4221.

The Kelvin River.   We must not fail to see the celebrated river. It is actually below the foregoing buildings (down a broad flight of steps) in Kelvingrove Park, Glasgow's prime outdoor summer centre. There are two large statues on the university side, one of Kelvin and one of the surgeon Joseph Lister, who began to develop his antiseptic surgical procedures in Glasgow before he moved to Edinburgh.

Strathclyde University.   Glasgow is privileged to have two major universities, both with considerable historic interest. The second one is Strathclyde University, chartered in 1964, but founded much earlier (in 1796) as Anderson's Institution. It is named for John Anderson (1726–1796), who had been a professor of natural philosophy at Glasgow University and had left instructions in his will that an institution be established "for the good of mankind and the improvement of science — a place of useful learning". He had no money to give towards this goal, but donated his library of 1500 books and his collection of scientific instruments. Private citizens of Glasgow provided the necessary cash and Anderson's wishes have been followed to this day: 60% of the current students at Strathclyde study science or engineering. The university has an impressive modern campus with a fine library — the Andersonian Library — named after the founder.

One of many distinguished Andersonians was Thomas Graham (1805–1869), a Glasgow native and an Anderson professor for seven years, after which he moved to University College in London. Strathclyde's Chemistry building is named after him; his portrait hangs in the large lecture room; there is also a statue.

Graham was a brilliant and inventive physical chemist, long before the subject "physical chemistry" was recognised as such, who merits greater recognition than he has received. His main interest was in diffusion ("Graham's Law"), culminating in his work on diffusion through membranes. He was the first to appreciate that many membranes are semi-permeable, blocking some substances while allowing others to pass freely, and to apply this property to the separation of dissolved substances from each other, a process he named "dialysis". He realised that particle size must be the basis for the separation, so that some substances (those that are blocked) must exist in solution as extremely large particles. He called these substances "colloids", after the Greek word for glue, and that gave rise to today's vigorous field of colloid chemistry. Dialysis is one of today's vital laboratory tools for biochemical separations, and especially invaluable in the dialysis machines that give life to thousands of people whose kidneys have failed them. Graham was the first president of Britain's Chemical Society (founded 1841). In 1854 he was appointed Master of the Mint, the last person ever to hold that title; Isaac Newton had been the first.

Strathclyde University occupies about 40 buildings on the east side of the city centre, between George Square and Glasgow Cathedral. The Graham Building and the Library are on Cathedral Sreet. The Royal College, an imposing building on George Street, has a fine statue of James Watt in the foyer. The university has a helpful Public Relations Service (McCance Building, corner George and Montrose Streets). Phone: (0141)-552-4400 ext 2196.

GREENOCK (Strathclyde)

Greenock, on the Firth of Clyde, was until recently a major shipbuilding centre. It is also the birthplace of James Watt, born here in 1736, and the city takes immense pride in the fact. The site of the house where he was born is downtown, close to the dominant *Victoria Tower*, and marked by a commemorative inscription. It is now used for local government activities, including "James Watt College". Further to the west is a memorial building (erected 1835 by Watt's son), which houses the Watt Monument Library, devoted exclusively to Scottish and local history — a rarely accessible source of such information, it might be noted. Next door (and in the same style) is the McLean museum, an all-purpose museum which highlighted a temporary exhibit "James Watt and the Age of Steam" when we visited, but always has some Watt material on permanent display. (The

Statue marking the birthplace of James Watt. Another statue stands in the town library.

temporary exhibit revealed a relatively litle-known fact about Watt: he was not single-mindedly focused on "steam", but invented and patented quite unrelated machinery, notably a couple of duplicating machines.)

Perhaps the most impressive item is a statue of Watt in the library, commissioned by public subscription. The inscription on it is a model of how civic pride should be expressed: "... not to extend a fame already identified with the miracles of steam, but to testify the pride and reverence with which he is remembered in the place of his nativity, ..."

The birthhouse is at the corner of William and Dalrymple Streets. The library and museum are on Union Street (corner Kelly Street). The museum is open Mon–Sat 10–12 and 1–5; the library is open 10–1 and 2–5 on Tue and Fri, 2–5 and 6–8 on Mon and Thu, 10–1 only on Wed and Sat. Phone: Information Centre in the Municipal Building, (01475)-24400.

INVERNESS (Highland)

Here we encounter military medicine at Culloden, a grand illustration of how science has been an intrinsic part of the fabric of British history. The battle of Culloden in 1746, which ended the Jacobite rebellion and the Stuart hopes of regaining the throne, was a particularly brutal one, characterised by many examples of what we would now classify as "atrocities". In attendance as physician to the national forces (containing loyal Scots troops as well as English) was Scotsman Sir John Pringle (1707–1782), born near Kelso in southern Scotland. He is the author of a classic treatise, *Diseases of the Army*, first published in 1752 and undoubtedly influenced by the infections that ravaged the army at Culloden and other battles. He was a pioneering spokesman for the provision of field hospitals with immunity from military action; he is rightly regarded as the founder of military medicine.

The point to be emphasised about this is that Pringle was no ordinary doctor. He was also professor of "pneumatics and moral philosophy" at Edinburgh University and renowned throughout Britain as a "basic" scientist. He was a close friend of Benjamin Franklin and Joseph Priestley and was president of the Royal Society in London from

1772 to 1778. From all that has been written about him, it is clear that Pringle was as humane a person as you could find — what he thought of the brutal inhumanity of the battle is not recorded.

The Culloden battlefield, about 5 miles east of the city, is a major tourist attraction, with a huge visitor centre and shop, open daily all year from 9 or 9.30 till early evening.

KINCARDINE-ON-FORTH (Fife)

James Dewar (1842–1923), born in Kincardine, was a versatile chemist, professor at the Royal Institution in London from 1877 until his death. His outstanding work was on the achievement of very low temperatures, where gases could be studied in their liquid state and in the course of it he invented the vacuum flask (also called "Thermos" or "Dewar" flask). This was not as simple as it may seem, for even a tiny amount of residual air in the wall of the flask will quickly equalize inside and outside temperatures. Ordinary vacuum pumps achieve only a _very low_ air pressure, but nothing like a true vacuum, so that they are not sufficient. Dewar had the expertise to solve the problem: charcoal (first cooled to −185°C) was found able to absorb the air remaining after a vacuum pump has done its work.

Kincardine remains to this day just a little town, seemingly out of place amidst the heavy traffic using the Kincardine bridge across the Forth. Dewar was born in a house that had been an old coaching inn before his time, and has become an inn again, the Unicorn Hotel, the town's major hostelry. A plaque identifies it as Dewar's birthplace, with appropriate words about his achievements. The hotel has a modern look, but all around are well-preserved old houses that give a good idea of how it must have been in Dewar's time.

Kincardine is on the north side of the bridge (A876) and it is easy to miss the turn off the main road. The Unicorn Hotel is on Excise Street, just a few steps from the High Street.

KIRRIEMUIR (Tayside)

Geologist Charles Lyell was born on the grand estate of Kinnordy on the outskirts of Kirriemuir. The family moved to southern England when Lyell was only one year old, but his father eventually came back to spend his old age here and Charles Lyell himself visited frequently and made several geological field trips in the surrounding area. The estate (with manor house intact) is now a wildlife refuge of the Royal Society for the Protection of Birds, but the present Lady Lyell still resides on the outskirts.

Lyell was here in 1848 (his father was then 81 years old) when a letter arrived to tell him that he had been recommended for a knighthood and that Queen Victoria had approved and was ready to confer this honour upon him. Originally, of course, knighthood had been for soldiers and a man would be dubbed a knight on the field of battle as reward for some feat of bravery — in fact, there was a special term of contempt, "carpet knight", for those who received the honour in the royal chambers. But by the 19th century the award was for any meritorious service to the nation, though the sword was still used for the dubbing ceremony, as it still is today. When Charles Lyell received his letter, Queen Victoria was at Balmoral Castle, 30 miles north of Kirriemuir, and Lyell rode there on horseback to receive the award in a personal ceremony just for himself — up Glen Clova along the southern Esk river, across the high hills around Lochnagar, and then down to the castle through Balmoral Forest. We can retrace at least half of this journey comfortably for ourselves, on the paved road (B955) which follows Glen Clova and terminates at a youth hostel. Beyond that there are only bridle trails. (It should be noted that Glen Clova, an impressive narrow valley with steep rocky sides, was one of Lyell's favourite field trip destinations in the area.)

The RSPB refuge (Loch of Kinnordy) is open all year round. Phone (01575)-572665 (Lyell estate) or (01575)-574097 (tourist office, summers only).

Netherhall, Lord Kelvin's former summer home, forms the nucleus of a new housing estate.

LARGS (Strathclyde)

Lord Kelvin took a turn to extravagance in his middle years and built himself a grand mansion just outside Largs, and bought a grand sailing ship to sail the waters off the coast. The house and grounds are now being developed into a housing estate — the house itself, we are told, will not be destroyed, but will be divided into apartments. The gate house also remains intact, with the adjacent iron gate and arch proclaiming the name "Netherhall". A blue plaque attests to Kelvin's residence here.

Netherhall is on the main coastal road, at the north edge of Largs.

MONTROSE (Tayside)

Robert Brown (1773–1858), a native of this town, was an outstanding botanist of his period, dubbed "botanicorum facile princeps" by none other than Alexander von Humboldt, and regarded by many as intellectually superior

295

to Linnaeus and others of his predecessors. He collected and classified plants from Australia and elsewhere; he was a protegé of Joseph Banks and became librarian and curator of the latter's huge accumulation of specimens; he was instrumental in donating this collection to the British Museum, which led to the creation of the museum's natural history collection, the nucleus for today's Museum of Natural History in London.

Uniquely among botanists, Brown's name is a household word for physicists and physical chemists. In 1828 he observed and described the spontaneous erratic motion of tiny pollen grains under the microscope, a phenomenon that became known as "Brownian motion". He thought at first that he had discovered the "primitive molecule" of living matter, the ultimate pulse of animate life, but later found that inorganic particles do the same thing. What can be the cause? Brown did not have the answer, but we now know that it is a manifestation of the general phenomenon of heat as molecular motion — pollen grains jiggle around as the result of impacts with myriads of smaller molecules around them. The explanation was the subject of one of Albert Einstein's famous 1905 papers, often regarded as a crucial paper for the acceptance of the concept of "molecules" by the scientific community, many of whom were still sceptical at the time because no one had ever actually seen a molecule.

Brown's work was done mainly in London, but Montrose is proud to list him among its handful of illustrious natives, in literature handed out to schoolchildren at the town's museum, for example. The Montrose public library is on the site of the house where Brown was born and it has a bust of him, with appropriate inscription, in the vestibule. (The museum's displayed collection, however, contains nothing explicitly about Brown.)

The library is on Mill Road, close to the railway station. It is currently undergoing renovation and the Brown bust may be moved upstairs as a consequence. The museum is nearby (opposite side of the High Street), at the corner of Museum Street and Panmure Place. Both are open daily except Sun. Phone (museum) (01674)-673232.

MUIR OF ORD (Highland)

Roderick Murchison (1792–1871), one of the best known geologists of his period was born at Tarradale House, seat of a long-established Highland family. The house has been much added to since his time and now belongs to the University of Aberdeen, who use it as a centre for field studies in geography and geology. Murchison's own work was done largely in Wales and the English/Welsh border area (*see* Cambrian Rocks, Wales p.258, and Silurian Rocks, Midlands (North) p.185), but he loved controversy and became involved (often quite unpleasantly) in all questions relevant to the definition of British rock strata. In later years he turned to geography and rallied support for many expeditions. The Murchison Falls of the Nile in Uganda are named after him. (More tourists undoubtedly see the falls than ever come to Muir of Ord.)

Muir of Ord is about 15 miles northwest of Inverness. Tarradale House is reached by driving 1.5 miles east of Muir of Ord along the A832, where a right turn to the house is signposted. The house is not open to the public, nor does it have a plaque to indicate the connection with Murchison. Phone (field centre) (01463)-870266. There is tablet in Murchison's memory in St. Clement's Churchyard, Dingwall, 6 miles north of Muir of Ord.

PEEBLES (Borders)

The Chambers brothers, William (1800–1883) and Robert (1802–1871), are the most illustrious natives of the town of Peebles, which is situated on a pretty site along the River Tweed. Tweed Green, on the left bank, is common land where the local people traditionally have had the right to hang out their washing to dry, a right exercised to this day.

The Chambers brothers are best known, of course, as founders of the Chambers publishing empire, with its famous dictionaries, encyclopedias and the like. Robert, however, was also a homespun scientist of considerable influence, for he believed, well in advance of Darwin, in some sort of evolution, and expressed this idea in a logical fashion in a famous book, *Vestiges of the Natural History of Creation*, published in 1844. He himself contributed little in

297

the way of direct evidence, but started with the by then generally accepted notion that new species appear at different times in the geological fossil record. One of his own original points was actually theological: given the succession of species, it is *belittling* to God to think that he would each time have had to undertake a new task of creation — better to think that a mechanism for adaptive change was built in at the beginning. The book aroused huge opposition (Hugh Miller in Cromarty was one of most vociferous opponents), which he had anticipated, for he published it anonymously — his authorship was not acknowledged till after his death. There was also at least one enthusiast — *Vestiges* is said to have been the trigger for Alfred Wallace's evolutionary theories. Charles Darwin had read it, too, but was appalled at the unscientific level of Chambers's thoughts.

The Chambers brothers used some of the riches they acquired from publishing to create the Chambers Institution, an imposing building on the High Street with a large inner courtyard. Completed in 1859, it houses the city library, a museum, art gallery and civic offices. There is a plaque outside, but no specific exhibit of Chambers history or memorabilia — a few books on the subject can be found in the library. On the opposite side of the High Street, a passage leads to a bridge across a small burn (the Cuddy) and on to "Old Peebles", now a residential area. The first street on the left is Biggiesknowe and number 18 on that street is the unpretentious house where the Chambers brothers were born, identified by a plaque. The house is actually much larger than appears from the front, as can be seen by a rear view from the other side of the burn.

The Library in the Chambers Institution is open Mon–Fri, 9–7 (Wed to 5.30). Phone (01721)-720123.

PITLOCHRY (Tayside)

Pitlochry has a very interesting hydro-electric power station, with a visitor centre which explains how power generation from flowing water is achieved — beginning with refreshingly expert reference to the experiments of Oersted and Michael Faraday. A major tourist attraction is a salmon ladder (34 pools joined by ascending steps), which allows salmon to move upstream past the power dam to their spawning grounds; windows allow visitors to see them below the water surface. The sight serves to remind us that there are still unsolved mysteries in science — molecular biology does not as yet have any idea of how fish manage to find their way to return unerringly from the distant oceans to their place of birth.

The Pitlochry power station is north of the city centre. It is open April to October, daily from 9.40–5.30. Between 5000 and 6000 salmon ascend the fish ladder each year, most of them during June to August. Phone (01796)-473152 or Pitlochry tourist office (01796)-472215.

Schiehallion (Loch Tummel). The road from Pitlochry along the north shores of Loch Tummel and Loch Rannoch provides some of the finest scenery in Scotland. The view to the south is dominated by the cone-shaped peak of Schiehallion (1083 m, 3547 ft). This mountain was used in 1774 by the then astronomer royal, Nevil Maskelyne, in a famous experiment to determine the density of the earth, but equally significant for demonstrating unambiguously that terrestial gravitation can act horizontally as well as towards the centre of the earth — it is a force that exists (as Newton assumed) between any two bits of matter.

Why did Maskelyne come all the way up here to do his experiment? Schiehallion's symmetry was the reason. He wanted to use a plumb line as his experimental tool — the mountain's mass should generate a deviation from the true vertical, from which the attraction for the lead weight can be calculated. But "true vertical" is hard to define and a symmetrical mountain allows one to make the measurement on opposite sides, where the force would be in opposite directions and "true vertical" itself need not be measured at all. Maskelyne, as we have said, was an astronomer — he used a telescope for the actual measurement, based on positions of distant stars relative to the plumb line.

Pitlochry and other places in this area lie on loop roads linked to the major highway, the A9. The road to Loch Tummel and Loch Rannoch (B8019) turns off the loop road between Pitlochry and Killiecrankie. There is a car park and visitor centre at Queen's View on Loch Tummel; a closer view of Schiehallion is obtained across the eastern end of Loch Rannoch. It is worthwhile to drive beyond Loch Rannoch (Rannoch Station, 38 miles from Pitlochry) where the terrain becomes glacial, with erratic boulders and glacial debris all around.

SICCAR POINT (near Cockburnspath, Borders)

James Hutton studied geological formations in many parts of Scotland in the course of developing his *Theory of the Earth*. Siccar Point, 35 miles south of Edinburgh, has the most striking and most easily accessible example of "unconformity", as well as providing a fine panorama of several miles of coastline. We can see multiple horizontal layers of pink sandstone sediment here, near the water line and partly washed away by the sea, and vertical strata of harder rock underneath — in many places the *vertical* strata thrust through the overlying sandstone. In no way could this formation represent sequential sediments from the sea. The vertical strata must be intrusions from deep in the earth, thrust into the softer rocks after melting, recrystallisation, and folding, illustrative of Hutton's perpetual cycling of the earth's crust.

Hutton was in an off-shore rowboat when he found this showpiece in the rock, in the company of John Playfair and Sir James Hall. The latter made a much copied sketch, and Playfair wrote eloquent prose about the occasion: "On us who saw these phenomena for the first time, the impression made will not easily be forgotten."

Siccar Point is about a mile south of the village of Cockburnspath and is approached through "Old Cambus Quarry", no longer a quarry, but a group of warehouses with huge parking space for trucks — the access road to the quarry (off the A1107) is sign-posted. To go to the point we proceed on foot from the parking area in a roughly northeasterly direction, through a wooden gate, up a hill to the left, and then across a field to the tops of the cliffs where the "unconformity" can be seen directly below, at the base of the cliffs. Total distance walked is about 300 yards. The quarry gates are open only during business hours, but it adds no more than half a mile to the walking distance if the car is left at the gate.

ST. ANDREWS (Fife)

Most people associate St. Andrews with the game of golf and one of the finest courses in the country, but it also has a claim to fame as a seat of learning all the way back to the 15th century. The University, founded in 1410, was the first in Scotland and the third in Great Britain (after Oxford and Cambridge). Buildings of two of the present colleges date from those early days and one of them, St. Mary's College, was home to James Gregory (1638–1675), member of a distinguished and scholarly Aberdeenshire family and inventor of the first reflecting telescope. Gregory was primarily a mathematician and his historical feat was to describe the _principles_ involved in constructing such a telescope. Unfortunately, he had (by his own admission) little practical skill and could find no optician capable of producing the lenses and mirrors he needed. The first usable reflecting telescope based on Gregory's design and calculations was actually built some years later by none other than Isaac Newton. The Upper Hall of St. Mary's College, a galleried room with pine panelling, much admired by Dr Johnson, was Gregory's workroom. The Senate Room on the first floor (built much later) contains two clocks flanking the fireplace, which were part of Gregory's scientific equipment.

St. Mary's College is on South Street, almost directly across from Holy Trinity Church.

10
Scottish Islands

ARRAN

The Isle of Arran is one of Europe's classic geological locations and geology students flock here for field trips from all over the world. An outstanding feature strikes the visitor from a distance, from the ferry that brings him to the island from Ardrossan on the mainland. It is the difference between north and south — the north half dominated by a huge mass of granite with the highest peak, Goatfell, rising to 2866 feet (874 m), the southern half flatter, more like the Scottish lowlands. It takes little imagination to see the northern mountains as an intrusion of igneous rocks, thrust up through sedimentary layers from pressures beneath the earth's crust. In fact, it was James Hutton who first postulated that this and other granite masses in Scotland were formed in this way and thereby created a new theory of the earth, sweeping aside all previous notions of the earth's age and history (*see* GEOLOGY p.31). Hutton visited Arran in 1787.

An adjunct of any Huttonian upthrust is that it should be surrounded by a grey "skirt" (as Hutton called it) of sedimentary layers which the upthrust has folded back into a steeply inclined position. This is less easy to see here on a grand scale, for successive ice ages in geologically recent times have obliterated much of the detail of the terrain and rich vegetation has covered most of it. One place where it is fairly clear is at Newton Point, seen across a little bay at the village of Lochranza at the north edge of the island. This is the place that Hutton himself cited as the example par excellence of his ideas. (One can walk to the point itself for a closer view.)

What makes Arran especially attractive is that a whole spectrum of geological phenomena can be seen here in addition to the foregoing. For example, so-called "erratic boulders" litter the shore — rocks carried by ice-age glaciers from their parent formations to places far away, where they are clearly in an anomalous environment. The beaches are also a good place to see small occurrences of steeply inclined sedimentary layers, made evident here by lack of vegetation. All this and much more is explained in exhibits in the island's Heritage Museum and in books and pamphlets that are for sale there — *Isle of Arran Trails: Geology* is recommended as an inexpensive and clear guide to geological features that can be reached by short walks. (The museum includes a restored "Smiddy" and other small buildings of general historical interest.)

Ferries between Ardrossan and the Arran port of Brodick run several times a day, with a crossing time of 55 minutes. The museum is about a mile north of the ferry terminus. It is open May to September, Mon–Sat 10–5. Phone: (01770)-302636.

GREAT CUMBRAE ISLAND (near Largs, Strathclyde)

Oceanography and marine biology, popularised today by the books and films of Jacques Cousteau and others, are fields of science that were first created by John Murray (1841–1914), an indefatigable explorer of the oceans and the sediments on their bottoms. He was born in Canada, to Scottish parents who had migrated there in 1834, but he returned to the land of his fathers as a young man, and did all his work here. And remarkable work it was, as he himself lacked any formal academic background. His practical education began as a scientific assistant on a three and a half year expedition aboard *H.M.S. Challenger*, where he took on the task of dredging and studying ocean bottoms. On his return he founded two marine stations, one in the Firth of Forth, which was short-lived, and a second (in 1886) here in the Firth of Clyde on the Isle of Cumbrae, which still flourishes, though now mainly as a teaching institution. For historical perspective, note that the two world-famous American institutions for oceanic research, Scripps in California and Woods Hole in Massachusetts, were founded considerably later, in 1901 and 1930, respectively. (Murray's work and publications, it is worth noting, were not confined to the oceans; he also organised and eventually completed by himself a survey of the _freshwater_ lochs of Scotland, the results of which were published in a 6-volume report.)

The present University Marine Biological Station is part of two university systems: London and Glasgow. The sites open to visitors are the Robertson museum and aquarium, named after David Robertson, whose interest and (in part) sponsorship were responsible for Murray's choice of the Cumbrae location for his station. The aquarium contains species native to the region; the museum blends the old and the new, with a good educational display on the planktonic basis of marine life, and historic displays about the early days of the station, when a wooden boat, the _Ark_, functioned both as laboratory and tight living quarters for the scientists.

Cumbrae Island is reached by ferry from the mainland town of Largs. Ferries operate frequently; the crossing time is only 10 minutes. Connecting buses take visitors to the marine station and the resort town of Millport. (The ferries take vehicles, but there is no need for a car except for visitors who want to make an extended stay on the island.) The museum is open Mon–Fri 9.30–12 and 2–4.45 (4.15 on Friday); also Easter Sat and all Sat from June to September. Phone (01745)-530581.

LEWIS AND HARRIS

The joined islands of Lewis and Harris are the most northerly of the Outer Hebrides. They are often visited for their isolation and scenery or, in the case of Harris, because it is the home of famous Harris Tweed. A tour of Lewis and Harris can be combined with tours of other islands, such as the Isle of Skye or the more remote islands in the Outer Hebrides chain.

When in Lewis and Harris for any reason, a visit to the

stone circle at Callanish should not be missed. Like the stone circles on Orkney it engenders a sense of mystery and seems to invite speculation. Who lived here? What have I in common with these forebears of mine? Physically, however, Callanish bears little resemblance to the Orkney circles. It is much smaller (diameter only 13 yards), contains a chambered tomb *within* the circle, and has impressive long stone-lined avenues leading up to it. A huge stone stands over the burial site, but it is not at the circle's true centre. It has been asserted by some that Callanish may have been another lunar and stellar site, capable of refined astronomical observations; others have envisaged it as a sacred ceremonial site. Professional archaeologists have been taciturn, seeing it as an enigma not easily solved. Enjoy it and add your musings to theirs!

Callanish is off the A858, about 10 miles west of Stornoway, the island capital and terminus of the principal ferry from the mainland (Ullapool).

ORKNEY ISLANDS

The Orkney islands are a fascinating place to visit, with unique physical beauty, and quite different in terrain, people and history from the Scottish mainland just a few miles away across Pentland Firth. The principal attraction for visitors are ancient monuments, going back to 3000 B.C. and even earlier. About half of them are on the main island and the rest are scattered over the smaller islands of the Orkney group. An added dividend to an Orkney visit is the magnificent bird life that can be seen along the rocky coastline. This is a breeding ground for kittiwakes, puffins, gulls, terns; all types of sea birds can be seen nesting in the cliffs in early summer. Use caution in approaching nesting terns — they tend to be nasty, swooping and diving at intruders.

Ring of Brogar and Standing Stones of Stenness. These two gaunt monuments are on the main island, less than a mile apart, not far from the town of Stromness. The Ring of Brogar is large, 114 yards in diameter (compared to 100 yards at Stonehenge), and most of its original 60 stones remain intact. Astronomical calendar advocates believe that Brogar was a lunar observatory, and an external stone, called the comet stone, 150 yards to the southeast, has been proposed as the viewing stone, from which the changing position of the moon could be observed. The standing stones of Stenness are the remains of a smaller circle, only 34 yards in diameter. Radiocarbon dating of animal bones or wood from holes or ditches associated with the circles suggest that Stenness is older than Brogar by perhaps 500 years. The viewing stone proposed by observatory enthusiasts in this case (a very tall stone called the watch stone) is directly north of the circle, consistent with the possibility that Stenness might have been used to observe the seasonal changes in the position of the sun.

To the untutored observer (like the authors of this book) the huge differences in geometrical dimensions and compass orientations among the stone circles generate scepticism about the idea that astronomical observations were the primary purpose of these structures. It's not just that Brogar and Stenness are different, but also Stonehenge in England and Callanish on the island of Lewis — no two of them seem to be alike. Another question, given the huge investment of labour needed for them: why erect the ring of Brogar when the stones of Stenness were already standing there, less than a mile away, just as they do now?

The Ring of Brogar on the island of Orkney

Whatever their true function, stone age people certainly went to a lot of trouble to erect these places. Some of the stones weigh close to 100 tons and were moved to their present location from quite large distances away. Presumably mechanical devices had to be invented to set them in upright positions.

Other neolithic monuments. Orkney has many other stone age monuments. Maes Howe, described as the "supreme example of a neolithic chambered tomb in Great Britain" is just a few hundred yards from Stenness. Archaeologists have assigned it the same approximate date as the standing stones, close to 3000 B.C. The excavated settlement of Skara Brae, on the Orkney coast about 6 miles to the northwest, dates from the same period. There was a thriving community here, contemporary with the early great dynasties of Egypt, but, as far as is known, in no way connected with them.

Rousay. Of the many outlying islands connected to Orkney by boat, Rousay has the most spectacular monument — the Midhowe chambered tomb on the island is very large and exceptionally well presented for public viewing. The chamber is over 100 feet long and 40 feet wide and is divided by upright stone slabs into orderly compartments, with a passage down the middle — tall enough for whoever entered (other than the dead, that is) to walk upright! Both skeletons and tools have been found here.

Norse settlements. Not all the sites worth visiting stem from the Stone Age. The Orkneys were settled by Norsemen around 800 A.D., and some of the monuments are from that period. The best are on the tidal island of Birsay, accessible from the mainland by foot at low tide, but cut off by the sea at high tide. Excavations here have revealed the foundations of a cathedral, bishop's palace and lesser dwellings. The Orkneys actually remained Norwegian territory — not part of Scotland at all — until 1468. St. Magnus Cathedral and other buildings in the island capital of Kirkwall were built by Norwegian rulers.

    If the crossing to Birsay is made just as the tide is going out, small tide pools on the rocks provide a rich source of sea anemones and other marine invertebrates.

Car-carrying ferries operate on regular schedule between Stromness on Orkney and Scrabster (near Thurso) on the mainland. Enquire locally about ferries to Rousay and other islands.

SHETLAND

The northernmost island of Shetland, Unst, became in 1819 the site of a uniquely French episode in science. After the revolution in France the new government established committees to deal with standards in weight and measurement and one of their most far-reaching decisions was to define a standard "metre", which they hoped would become (as indeed it virtually has) the universally accepted international unit of length. Their arbitrary choice was seemingly simple — 10^{-7} of a quadrant of the earth (one quarter of the circumference) on the line that passes through the city of Paris. Unfortunately this terrestial distance was not accurately known, so that a number of distinguished French scientists were set the task of determining it precisely, among them Jean Baptiste Biot and Dominique Arago. A complicating factor is that the earth is not exactly spherical, so that the deviation from sphericity had to be estimated at appropriate places, which Biot and Arago did by a time-honoured Newtonian method, measuring the time constant of the swing of a pendulum as a function of the length of the pendulum rod, from which the force of gravity at each point can be computed.

The two physicists travelled north from Paris with their equipment, along the chosen pathway, until they came to Unst, the end of all terra firma in that direction. They then returned to Paris and their final act was a prosaic one, to make two scratch marks on a metal bar to indicate — indelibly, but not terribly accurately — the metre length decreed by the legislation. This bar was then kept at Sevres near Paris and everybody else came from all over the world to calibrate their own standards of length in terms of the French metre. Strange as it may seem, it was not until the second half of the 20th century that a more precise and less cumbersome definition of the metre, based on spectroscopic measurement, was adopted.

Biot and Arago did their work at Buness House at Baltasound on Unst. There is reportedly a memorial stone to recall their visit, but we (more timid than they were) have not yet braved these northern latitudes to inspect it.

STAFFA (uninhabited island, off the island of Mull)

Huge impressive basaltic columns are the attraction here, textbook examples for geologists. They were formed in an era of violent volcanic action, along a line that can be traced from the Island of Skye to the north of Ireland (*see* GIANT'S CAUSEWAY, Ireland p.323). The most spectacular are in Fingal's Cave, all sides of which (roof and ground included) are composed of crystalline pentagonal or hexagonal pillars, with uniformly spaced transverse joints. The caves were first brought to the attention of the outside world by Joseph Banks, who was unintentionally driven into the area by unfavourable winds on the way home from a journey to Iceland. Their volcanic origin was initially disputed by champions of the literal truth of the bibilical story of Noah's flood, who speculated that the columns crystallised *in situ* from dissolved salts as the flood waters receded.

Fingal's Cave is of course well known to music lovers as the inspiration for music by Mendelssohn. Much as we admire the latter, we don't think it does justice to this dazzling site.

Boat trips around Staffa originate from the island of Mull, from April to September. Exploration of Fingal's Cave on foot is possible in good weather. Phone (Tourist office at Tobermory) (01688)-2182.

11
Ireland

Ireland is dearly loved by literary tourists. In Dublin they can retrace the steps of Leopold Bloom on 16 June 1904, as recounted in James Joyce's *Ulysses*, or they can visit the haunts of James Joyce himself, as described in his autobiographical works. Up north in County Sligo they can make a pilgrimage to the Lake Isle of Innisfree and other sites associated with the poet William Butler Yeats. But there is much scientific history, too. In reading about it we must appreciate that Ireland as a separate nation is a modern concept and that Irish science was part of British science before then. For example, the fledgling British Association for the Advancement of Science (founded in York in 1831) met in Dublin in 1835. The year before it had met in Edinburgh, the year after it met in Bristol.

The most famous scientific native of the country was Robert Boyle (1626–1691), the "sceptical chymist" and one of the leading lights in the founding of the Royal Society in London (*see* CHEMISTRY p.41). He was the youngest son of the first Earl of Cork, an Anglo-Irishman who acquired wealth and lands in Ireland as an Elizabethan immigrant. The Earl sent Robert off to Eton and to the continent of Europe for his education and Robert never returned. Needless to say, Robert left no marks on his native soil, but, as if to compensate, his father the Earl left plenty — they show us an often coarse and vain character, food for thought for all who like to ponder parental influences in the creation of intellectual genius.

BIRR (County Offaly)

William Parsons, Earl of Rosse (1800–1867) built what was then the world's largest telescope on his estate at Birr in central Ireland (improving on the best of the Herschel telescopes), and for 75 years astronomers came here from every land to see further into space than was possible at home. With this instrument, Parsons himself became the first astronomer to see a spiral nebula, the kind that make such spectacular pictures in elementary textbooks and are now known to be galaxies, much like our own galaxy. Parsons's achievements are truly remarkable because he conscientiously devoted most of his energies to his duties as lord of the manor, and he built the telescope entirely by himself with the aid of only local labour.

Birr Castle sits is the midst of a 100-acre demesne, one of the most attractive grounds in all of Ireland. It includes a bit of the Little Brosna river, waterfalls, a lake, formal gardens, an arboretum, etc. — and, on an open lawn, the remains of the telescope, firmly anchored to two huge gothic stone walls. Some fine pictures of nebulae are exhibited in glass cases on the walls. The castle itself is still the private residence of the Earls of Rosse (it has in fact been in the Parsons family since the 1620s), but the grounds are open to visitors every day of the year.

CHARLEVILLE (County Cork)

Most readers will know what an orrery is – an ingenious device, driven by clockwork, for accurate display of the relative motions of the earth, moon and planets around the sun, designed to demonstrate the change in seasons, eclipses, and the vagaries of planetary positions as seen from the earth. The origin of the word "orrery", however, is less well known. It happens to reside here in Charleville, a town in the middle of the barony of Orrery and Kilmore, a former subdivision of County Cork, no longer extant in today's organisation of county government. Roger Boyle, third son of the first Earl of Cork, a more flamboyant man than his younger brother Robert, was given the barony and created the first Earl of Orrery in 1660. He founded the town, which he named Charleville for the newly restored King Charles II. Roger, though acknowledged as the source of the town's prosperity, was a cruelly repressive lord of the manor and is not remembered fondly. (Since about 1690 none of the Boyles has actually lived here, though the now combined earldom of Cork and Orrery still exists – the present titleholder, Patrick Reginald Boyle, is chronologically the 13th.)

The astronomical connection comes from Charles Boyle, grandson of Roger and 4th Earl of Orrery. He became fasciniated by the astronomical clock and had one of the finest early examples built for himself by clockmaker T. Rowley in 1712. The earl was such a force for popularisation of the instrument (among the London gentry) that his name became synonymous with it and has been ever since.

The original Rowley orrery can be seen at the Science Museum in London. Charleville is shown as "Rath Luirc" on many maps and some road signs, but this name is no longer used. The town today is almost entirely Catholic and the former Anglican church has been converted to the town library.

DUBLIN

Most science in Dublin has been associated with Trinity College, a venerable educational institution founded in the reign of Queen Elizabeth I. One of its first students (entering the college in 1594) was James Ussher (1581–1656), whose calculation of the age of the earth, based on biblical chronology, was enormously influential for over 100 years, a dogma that frustrated every geologist with independent ideas on the subject. Ussher's thesis — in which the origin of the world was dated as the night before 23 October in 4004 B.C. — was the result of meticulous calculations, based on the genealogies in the books of the Old Testament and supporting manuscripts in Greek, Aramaic and other languages. No one looked on this as a crackpot clerical tract. It was taken seriously as a product of admirable historical scholarship and hotly debated by other scholars. John Lightfoot, for example, noted Cambridge scholar and vice-chancellor of the university at the time, argued that the correct date and time of the creation should be taken as three days later than Ussher's date, namely 26 October 4004 B.C. at 9.00 a.m. Even Isaac Newton took Ussher's dates seriously. He estimated, in his *Principia*, on the basis of the earth's rate of cooling, that it must be at least 50 000 years old, but he was uncomfortable with the result and speculated about latent causes that might have accelerated the cooling process — he did not consider his estimate as a serious denial of Ussher's date.

(Ussher's career after he left Trinity veered to the Church and he became an archbishop. His chronologies were actually printed in the margins of bibles from 1701 onwards and well into the early 20th century, and they came to be regarded with the same reverence as the biblical text itself. Here are some of the "standard" dates: Noah's flood, 2349 B.C.; exodus from Egypt, 1491 B.C..; accession of King David, 1056 B.C.)

An example of a more recent Trinity scholar, 3 centuries later, is George FitzGerald (1851–1901), one of the so-called "Maxwellians", the small group of physicists who immersed themselves in Maxwell's theory of the electromagnetic field and amplified and extended it (*see* ELECTRICITY & MAGNETISM p.20). His most celebrated creation is the phenomenon later known as "FitzGerald's Contraction" — the at first seemingly absurd notion that objects contract as their speed of motion approaches the speed of light. It arose as a possible way out of the dilemma posed by American experimenters (Michelson and Morley) who had failed to find evidence for an "ether", the medium which was thought at the time to be theoretically necessary for explaining the charge movements of electromagnetic waves. The idea first emerged during a visit to FitzGerald's Liverpool friend Oliver Lodge, who later called it a "brilliant suggestion" that "flashed on" FitzGerald in the course of one of their innumerable discussions. FitzGerald himself did not think it was important at the time and most British physicists took little notice until after FitzGerald's death. But then it became known that the Dutch physicist Hendrick Lorentz had similar ideas about the same time and both he and FitzGerald are now seen to have been precocious geniuses, grasping for new concepts that emerged in full form a few years later in Einstein's Theory of Relativity.

Trinity College. The present buildings date from the mid-18th century. The most impressive is the Old Library — particularly the Long Room within it, with its high vaulted ceiling, all made of dark wood. The library's most treasured possession, the *Book of Kells* (mediaeval illuminated gospel) is kept here and some pages are always open for display. The central aisle is lined with busts of famous men, Trinity graduates among them. The graduates include Archbishop Ussher, William Hamilton (*see* below) and Jonathan Swift, author of *Gulliver's Travels*. Aristotle, Cicero, Newton, and the Duke of Wellington are some of the honoured outsiders.

The College Chapel and the Examination Hall are two much smaller buildings, facing each other across Parliament Square, and both contain tributes to William Ussher. His portrait hangs alongside portraits of Queen Elizabeth and Jonathan Swift in the Examination Hall. In the Chapel the central panel of a triptych of stained glass windows is dedicated to his memory. Bishop Berkeley, a famous Trinity-educated cleric a century later than Ussher, with profound views on our perception of nature, is honoured similarly to Ussher in both buildings.

A portrait of George FitzGerald hangs in the engineering school of Trinity College. The house where he was born (19 Lower Mount Street) is now a Catholic hostel. There is no plaque.

The Examination Hall is used for examinations, convocations, etc., and is not normally open to the public, though the staff will let you take a peek if it is not in use. The Ussher portrait here is actually a copy, the original being in the Provost's private residence.

St. Patrick's Cathedral This is Church of Ireland, on the south side of the river Liffey, the Catholic Cathedral being on the north side. The cathedral contains a huge number of memorials — a sort of Irish analogue of Westminster Abbey — including a 20th century window dedicated to William Ussher (a student at the cathedral choir and grammar school before he entered Trinity College) and the actual tomb of

Jonathan Swift. The latter was Dean of the cathedral and himself composed the provocative epitaph inscribed on his stone.

Of greatest relevance to this book is the monstrous and vulgar memorial to the family of Richard Boyle, Earl of Cork and father of Robert Boyle, which was erected by the Earl himself for the perpetual glorification of the Boyles. It is forty feet tall, formally a memorial to his second wife (the one who bore all his children), with effigies of her parents and grandfather, the Earl and Countess themselves (recumbent, it should be noted, he lying on top of her), and eleven of their children in a row at the base. The infant in the middle of this row is presumably Robert, born in 1627 just before the monument was erected. The artistry is execrable. All the children, including the infant, have old faces and only the figure size indicates chronological age. This tasteless structure stood originally next to the high altar, but objections were raised — should the congregation be expected to kneel and pray in front of the Earl of Cork and his wife? — and the monument was moved to the opposite end of the nave. (All of this happened in the Earl's lifetime and he protested vehemently.)

For relief from the vulgarity, look at the stairs to the organ loft. They are recent, but artistically in good taste.

Flash of genius at the Royal Canal. William Rowan Hamilton (1805–1865) was a mathematician at Trinity and chief astronomer at Dunsink Observatory. He was not somebody who turned the world upside down in his lifetime, but he had some brilliant mathematical ideas, which turned out 70 years later to be exactly what was needed by theoretical physicists to express new ideas in their field — ideas beyond the capacity of traditional "classical" mechanics to handle. One was the growing realisation that matter consists of myriads of particles in random motion, which led to the development of statistical mechanics. The other was the conceptual notion that energy transfer involves discrete "lumps" of energy called *quanta*, which gave rise to quantum mechanics. The two Hamilton innovations that solved the physicist's problems were (1) a novel general equation of motion for assemblies of many particles and (2) the inspired invention of a new kind of mathematical function

(of which vectors and tensors are special examples) called
quaternion.

Hamilton claimed that the formulas that essentially define
quaternions came to him in 1843 in a "flash of genius", as
he was walking with his wife on the towpath of the Royal
Canal. It happened where Brougham Bridge crosses the
canal at Broombridge Road in the suburb of Cabra, and
Hamilton stopped and scratched the formulas into the stone
of the bridge. It sounds like an apocryphal tale, but the evidence for it is strong — Hamilton apparently always had a
need to "see" formulas written out and was well known for
his habitual use of any available surface for his jottings. The
inscription has long since mouldered away, but there is a
plaque on the bridge commemorating the event, with the
crucial formula,

$$i^2 = j^2 = k^2 = ijk = -1$$

inscribed upon it. The plaque was put up at the instigation
of Eamon de Valera, Ireland's militant republican premier,
who had ambitions in his youth to be a physicist or
mathematician and took great pride in Hamilton's achievements.

Brougham Bridge is at what is now an ugly and ill-kempt segment of the Royal Canal in the suburb of Cabra. Bus routes 22
and 22A terminate near the bridge and provide an alternative
means of approach for those who don't want to take the long
towpath walk.

Merrion Square.   This is one of several attractive squares
in Dublin, formerly purely residential, now mostly devoted
to offices. There are plaques to show who lived where. On
the south side we have the "Liberator", Daniel O'Connell, at
number 58, W.B. Yeats at number 82, and, perhaps surprisingly, the Austrian theoretical physicist Erwin Schrödinger
at number 65. His presence in Dublin was a result of prime
minister de Valera's aforementioned interest in physics and
mathematics — de Valera set up an institute explicitly to
attract refugees from war-torn Germany — and the plaque
states that he worked in this house from 1940 to 1956. His
work was not limited to physics, for he also ventured into a
physicist's view of biology while he was here, gave lectures

on that subject at Trinity College, and published a book (*What is Life?*) which is said to have been influential in persuading other physicists to become involved in biological research. It is, however, reliably reported that Schrödinger was captivated by the Irish ladies and spent more time on affairs of the heart than on either physics or biology.

University College. Another physicist worthy of note is George Stoney (1826–1911), who invented the word *electron* for the fundamental unit of electric charge. He did not imply the existence of electric particles that carry a single unit of charge, but when J.J. Thomson discovered such particles in Cambridge the word *electron* was soon applied to them. (But not by Thomson himself — he just called them "corpuscles".) Stoney spent almost all his working life at what used to be called Queen's University, but later became Ireland's National Unversity. University College is its major component. It was originally located on Earlsfoot Terrace, south of St. Stephen's Green, but most of it is now in a modern campus in the suburb of Belfield, though some departments remain in the wings of the old building. The former Examination Hall at its centre has been converted into Dublin's splendid National Concert Hall.

LISMORE CASTLE (County Waterford)

Lismore Castle, to this day the most grandiose castle in all of Ireland, is where Robert Boyle was born in 1627. The castle was part of Richard Boyle's purchase from Sir Walter Raleigh. It became the property of the Dukes of Devonshire (the Cavendishes) in 1753 by marriage of a Cavendish to a daughter of the fourth Earl of Cork. The gardens are open to the public in the summer months, but the castle itself is a private residence. One can get a fine view of the castle all year round from the road that enters Lismore from Clogheen to the north — a highly scenic road, crossing high moorland, worth taking for its own sake. Another scenic route (quite different in character, mostly wooded), runs south from Lismore to Youghal. Part of the route runs alongside the Blackwater River, which was the main traffic vein of the Boyle "empire" in the olden days. Robert must have travelled it often in his youth. (There is yet another Boyle castle in Ballyduff, a few miles west of Lismore. This one was built by Richard Boyle himself in the 1620s.)

SKREEN (County Sligo)

George Stokes (1819–1903) was born here in the north-western part of Ireland, the son of the rector of the parish church. Somehow he found his way to a superior education and ultimately to Cambridge, where he graduated with the highest possible distinction (senior wrangler) and, a few years later, was appointed to Isaac Newton's old position of Lucasian professor, a post he held for more than 50 years! He became the leading authority on motion of small particles in fluids and his equations on that subject are still required knowledge for students in physics and even other fields. In biochemistry, for example, the "Stokes radius" is a common way to characterise molecular size, based on Stokes's equations — an almost unique parameter because it gives the size of single molecules *in solution*, the state in which many of them would be inside a living cell. Stokes also had an avid interest in optical phenomena. He was the first person to understand fluorescence (in 1855) and it was he who gave the phenomenon its name.

Stokes remained a humble man for all of his long life and retained the devout religious feelings of his childhood. He was deeply interested in the relation between science and religion and gave lectures on the topic which were published in the 1890s.

Skreen is a village about 12 miles southwest of Sligo, between Ox Mountain and the sea, on a turning off the N59. Plans are afoot to install a memorial to Stokes in the Anglican parish church. (When in this area, do not miss seeing beautiful Lough Gill, 2 miles southeast of Sligo. In the far corner of it is the tiny "Lake Isle of Innisfree", made famous by the poet W. B. Yeats.)

YOUGHAL (County Cork)

Youghal is a narrow ribbon of a town along the edge of the estuary of the Blackwater River, with walls against invaders

*Details of the Boyle monument in St. Mary's Church in Youghal.
The great Earl, recumbent, is above. Some of his 14 children are
below. Robert Boyle was not yet born when this monument was
erected, but he is included on a similar (equally tasteless) monument in St. Patrick's Cathedral in Dublin*

from the sea. The town was Sir Walter Raleigh's seat during his brief sojourn in Ireland before he sold his lands cheaply to Richard Boyle; the Elizabethan mansion where he lived is still a private residence on spacious grounds adjacent to St. Mary's Church. Richard Boyle quickly grew rich, became the first Earl of Cork and built the Youghal harbour to provide a link to England for his produce.

St. Mary's Church (originally dating from 1220) contains a large and vulgar Italian-style monument to the Boyles, erected by the Earl himself, with carvings of the Earl and both his wives, the second (Lady Katherine) in noble ermine, and nine of his fourteen children ranged along the base — but not yet Robert, then still unborn. The wrought iron railings in this part of the church bear a score of coats of arms of the Boyles and the families with which they intermarried.

Near the church is New College House, on the site of a medieval college which the Earl acquired with the rest of the land. He put it in charge of his brother (whom he had made Bishop of Cork) so that most of the collegiate revenues would end up in his own pocket. But let it be noted that the Earl had his generous side as well. He erected six substantial almshouses in 1610 and (as the plaque tells us) "provided £5 apiece yearly for each of ye six old decayed soldiers or Alms Men for ever." The terraced houses are still used for needy widows and were in the process of being converted into apartments when we visited.

In Northern Ireland

ARMAGH (County Armagh)

Armagh is the ecclesiastical capital of all of Ireland. Both Anglican and Catholic primates have their seats here and the two cathedrals (both named after St. Patrick) face each other across the roofs of the city. The city itself is steeped in history and surrounded by lush green Irish countryside.

The most celebrated of the churchmen is Richard Robinson, who was Anglican archbishop from 1765 to

1795. He had a keen interest in science and his noblest achievement was the founding of the Armagh observatory, an institution which remains to this day an active training and research centre. A planetarium for the general public and a "Hall of Astronomy" have been added in recent years, the latter being a modern style museum with video displays and other devices to intrigue and educate schoolchildren and other visitors. Robinson also founded a public library, across the street from the Anglican cathedral, which is beautifully maintained to perpetuate a genuine 18th century atmosphere. Noteworthy items in its collection (displayed in a glass case) are original editions of some of the works of James Ussher, famous for his determination of the age of the earth on the basis of biblical accounts (see DUBLIN). Ussher was formally Archbishop of Armagh from 1625 to 1656, but left Ireland at the time of the 1641 uprisings against English landlords and never returned to resume his duties.

St. Patrick's Church of Ireland Cathedral has a bust of Archbishop Robinson and commemorative windows for him and Archbishop Ussher. Stated opening times are unreliable, but admissioin can usually be obtained by ringing the outside bell at the public library, which is also how one gains access to the library itself. Phone (01861)-522611 (office of the verger). The planetarium and museum are open Mon–Fri, 10–5, Sat 1.30–5, Sun (April to August) 1.30–5. For times of planetarium shows phone (01861)–523689. (Public star-gazing sessions are held at the observatory on some winter evenings.)

BELFAST (County Antrim)

William Thomson (Lord Kelvin), Glasgow's great contribution to the world of science, was actually born in Belfast. His forebears were Scottish farmers, who had migrated to Ireland in the 1640s, but his father, James Thomson, broke the bond to the land. He was a man of remarkable ability and strong character, who taught himself science and mathematics at home and then went for four successive winters to Glasgow for formal studies and a degree. His reward was a post as teacher at the Royal Academical Institution and eventual appointment (in 1832) as professor of mathematics at the University of Glasgow — whereupon he and his children moved permanently to the Scottish city. William always remainied cognisant of his debt to his father. When he was installed as Chancellor of the University of Glasgow in 1892, for example, he gave an eloquent picture of the hardships of the annual sea journey from Belfast in his father's time.

The Thomson home in Belfast was at College Square East, but has been demolished. There is however a Kelvin statue at the Botanical Garden, on Stranmillis Road, close to Queens University and the Ulster Museum.

GIANT'S CAUSEWAY (near Bushmills, County Antrim)

"Ye cliffs and grots where boiling tempests wail". This is a spectacular volcanic site, similar to the one on the Island of Staffa in the Hebrides, but more easily accessible. The sharply sculptured basaltic columns were once the object of furious geological controversy: the Vulcanists saw them as crystallised products of hot molten rocks from the earth's interior; the Neptunists thought they were deposited from receding waters of a globe-encircling flood. The problem was settled in favour of the Vulcanists in the Auvergne in France, where more modest columns of a similar kind exist in the midst of incontrovertibly volcanic land.

Stones of Contention. Giant's Causeway, from a painting by Susanna Drury in the 1730s.

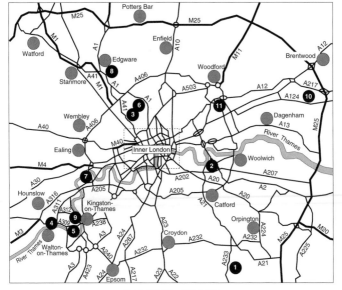

GREATER LONDON

The rectangle near the centre of the map is the area covered by the map for CENTRAL LONDON.

1. Downe
2. Greenwich Observatory
3. Hampstead
4. Hampton
5. Hampton Court
6. Highgate Cemetery
7. Kew Gardens
8. Mill Hill
9. Teddington
10. Upminster
11. Wanstead

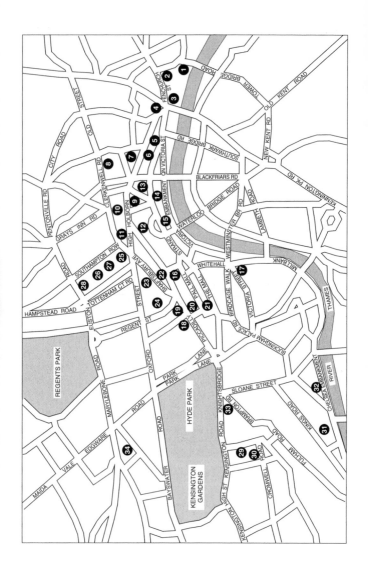

CENTRAL LONDON

1. Tower of London (Old Royal Mint)
2. St. Olav's Church
3. The Monument
4. Royal Exchange
5. St. Paul's Cathedral
6. Warwick Lane
7. St. Bartholomew's Hospital
8. Charterhouse
9. Barnard's Inn (Gresham College)
10. Gray's Inn
11. Red Lion Square
12. Lincoln's Inn Fields (Royal College of Surgeons)
13. Crane Court (Fleet Street)
14. The Temple
15. King's College (Strand)
16. National Portrait Gallery (Trafalgar Square)
17. Westminster Abbey
18. Royal Institution (Albemarle Street)
19. Burlington House (Piccadilly)
20. Jermyn Street
21. Royal Society (Carlton House Terrace)
22. Leicester Square
23. Soho Square
24. Broadwick Street (John Snow Pub)
25. British Museum
26. University College
27. London School of Hygiene & Tropical Medicine
28. Wellcome Institute
29. Science Museum
30. Natural History Museum
31. Carlyle's House (Cheyne Row)
32. Chelsea Physic Garden
33. Rutland Gate (South Kensington)
34. St. Mary's Hospital (Fleming Museum)

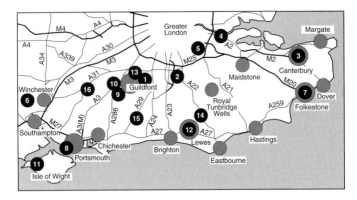

SOUTHEAST ENGLAND

1. Albury
2. Burstow (near Horley)
3. Canterbury
4. Dartford
5. Downe
6. East Tytherley (near Romsey)
7. Folkestone
8. Gosport

9. Haslemere
10. Hindhead
11. Isle of Wight
12. Lewes
13. Ockham
14. Piltdown (near Uckfield)
15. Pulborough
16. Selborne (near Alton)

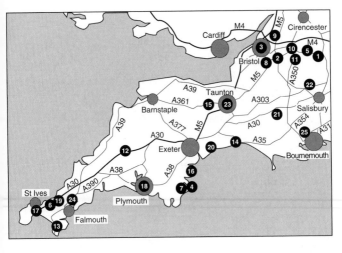

SOUTHWEST

1. Avebury
2. Bath
3. Bristol
4. Brixham
5. Calne
6. Camborne
7. Dartmouth
8. High Littleton
9. Horton (near Chipping Sodbury)
10. Kington St. Michael
    (near Chippenham)
11. Lacock Abbey
12. Launceston
13. Lizard
14. Lyme Regis
15. Milverton
16. Paignton
17. Penzance
18. Plymouth
19. Redruth
20. Sidmouth
21. Stalbridge (near Sherborne)
22. Stonehenge
23. Taunton
24. Truro
25. Wimborne Minster

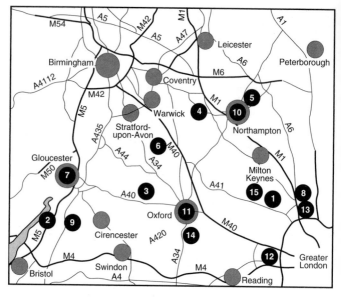

MIDLANDS (SOUTH)

1. Aldbury (near Berkhamsted)
2. Berkeley
3. Churchill (near Chipping Norton)
4. Daventry
5. Ecton
6. Edgehill
7. Gloucester
8. Harpenden
9. Minchinhampton
10. Northampton
11. Oxford
12. Slough
13. St. Albans
14. Sunningwell (near Abingdon)
15. Tring

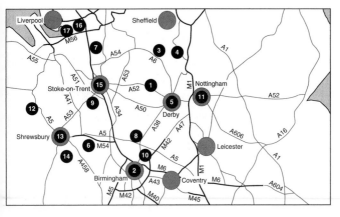

MIDLANDS (NORTH)

1. Ashbourne
2. Birmingham
3. Chatsworth
4. Chesterfield
5. Derby
6. Ironbridge (near Telford)
7. Jodrell Bank
8. Lichfield
9. Maer
10. Middleton Hall (near Birmingham)
11. Nottingham
12. Oswestry
13. Shrewsbury
14. Silurian Rocks
15. Stoke-on-Trent
16. Warrington
17. Widnes

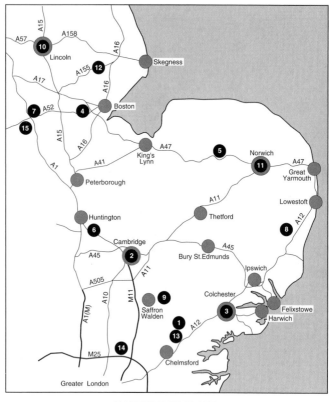

EASTERN ENGLAND

1. Black Notley (near Braintree)
2. Cambridge
3. Colchester
4. Donington
5. East Dereham
6. Godmanchester
7. Grantham
8. Halesworth
9. Hempstead
   (near Saffron Waldon)
10. Lincoln
11. Norwich
12. Revesby
13. Terling (near Chelmsford)
14. Waltham Abbey
15. Woolsthorpe (near Grantham)

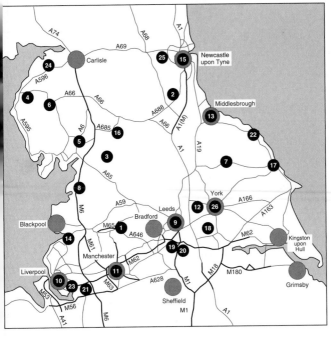

NORTHERN ENGLAND

1. Burnley
2. Byers Green
   (near Bishop Auckland)
3. Dent (near Sedbergh)
4. Eaglesfield (near Cockermouth)
5. Kendal
6. Keswick
7. Kirbymoorside
8. Lancaster
9. Leeds
10. Liverpool
11. Manchester
12. Marston Moor (near York)
13. Marton (near Middlesbrough)
14. Much Hoole (near Preston)
15. Newcastle
16. Outhgill (near Kirkby Stephen)
17. Scarborough
18. Selby
19. Thornhill (near Dewsbury)
20. Wakefield
21. Warrington
22. Whitby
23. Widnes
24. Wigton (near Carlisle)
25. Wylam (near Newcastle)
26. York

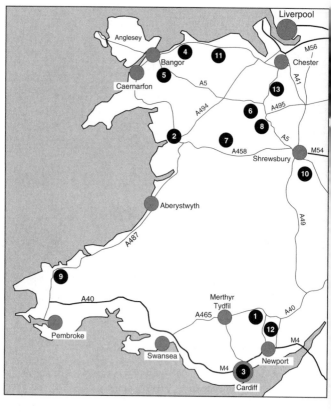

WALES

1. Blaenavon
2. Cambrian Rocks (Dolgellau)
3. Cardiff
4. Conwy
5. Cwm Idwal (near Bangor)
6. Glyn Ceiriog (near Llangollen)
7. Lake Vyrnwy
8. Oswestry
9. Preseli Hills (near Fishguard)
10. Silurian Rocks
11. St. Asaph
12. Usk
13. Wrexham

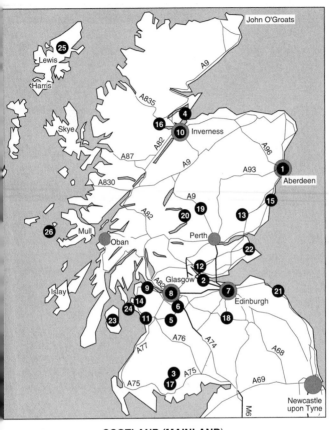

SCOTLAND (MAINLAND)

1. Aberdeen
2. Bo'ness
3. Corsock (near Dumfries)
4. Cromarty
5. Darvel (near Kilmarnock)
6. East Kilbride
7. Edinburgh
8. Glasgow
9. Greenock
10. Inverness
11. Irvine
12. Kincardine-on-Forth
13. Kirriemuir
14. Largs

15. Montrose
16. Muir of Ord
17. Parton (near Dumfries)
18. Peebles
19. Pitlochry
20. Schiehallion (near Pitlochry)
21. Siccar Point (near Cockburnspath)
22. St. Andrews

SCOTTISH ISLANDS

23. Arran
24. Great Cumbrae
25. Lewis and Harris
26. Staffa

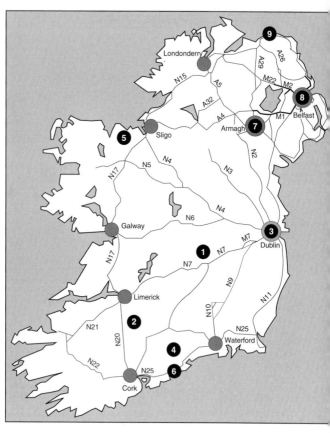

REPUBLIC OF IRELAND

1. Birr
2. Charleville (Rath Luirc)
3. Dublin
4. Lismore Castle
5. Skreen
6. Youghal

NORTHERN IRELAND

7. Armagh
8. Belfast
9. Giant's Causeway

INDEX OF NAMES

Names in *italics* represent political or literary figures.

INDEX OF PLACES

SUBJECT INDEX

References are to substantive text on the indexed topics.